F-51D MUSTANG

PART I—FLIGHT HANDBOOK, USAF SERIES F-51D AIRCRAFT

PART II—APPENDIX (follows flight handbook)

- World War II advertising
- Photographs showing development of Mustang
- Typical Surplus Sales offering

AVIATION BOOKS BY
FLYING ENTERPRISE PUBLICATIONS
3164 Whitehall • Dallas, Texas 75229

MITCH MAYBORN - Publisher

- Stearman Guidebook
- F-51D Mustang Handbook
- Cessna Guidebook

Early Military Aircraft of the First World War
- Volume I—Landplanes
- Volume II—Seaplanes & Motors
- Volume III—Airships

Flying Enterprise Publications
ISBN No. 0-912470-13-5

T.O. No. 1F-51D-1
(Formerly AN 01-60JE-1)

FLIGHT HANDBOOK

USAF SERIES
F-51D
AIRCRAFT

Commanders are responsible for bringing this handbook to the attention of all personnel cleared for operation of affected aircraft.

Published under authority of the Secretary of the Air Force and the Chief of the Bureau of Aeronautics.

This publication replaces T.O. No. 1F-51D-1 (formerly AN 01-60JE-1), dated 15 July 1952 and Safety of Flight Supplements thereto: -1C and -1D. This book was complete at time of issue, since there were no outstanding Safety of Flight Supplements.

20 JANUARY 1954

REISSUED: 1966
1971
1973

Table of Contents

Making a pass at an enemy tank is as much of a gamble as making a pass with the dice at Las Vegas, if you don't know your airplane. You can't change the odds in a dice game, but you *can* increase your odds of making a successful tank kill. Your knowledge and superior equipment will put the odds in your favor, provided you read about and understand the capabilities of your airplane as presented in your Flight Handbook. These first pages will let you know where you can find this information in your handbook. Learn to know your airplane by starting right here, for this handbook is the only technically accurate and constantly current source of F-51D operating data. The information in this handbook is based on engineering and flight test experience of the Air Force and the manufacturer, as well as the service experience of the using commands. North American Aviation and the Air Force have carefully considered your handbook requirements and have cooperated to prepare this handbook in a completely new style that definitely marks the old -1 T.O. obsolete. These new-type handbooks not only are more attractive, but are easier to read and easier to use. You'll note that full use is made of illustrations to highlight descriptions and specific procedures. The flight handbooks for all airplanes have not been prepared to the new specification, but the new books can readily be identified by the cover. The old-style book has a small, rectangular photo of the airplane centered on the cover; the new handbook has a full-page cover illustration.

This handbook was prepared solely for your benefit, and you as the pilot of an F-51D should make sure you have a copy for your own personal use. *Air Force Regulation 5-13* specifically provides that each pilot *(except those attached to an administrative base)* is entitled to his own copy of the Flight Handbook for his airplane. Don't let anyone tell you otherwise.

Once you have your handbook, take time to read and study it completely to gain an over-all knowledge of the airplane, and keep it handy for a reference guide.

The Technical Order distribution system works surprisingly well if you do your part. In this respect, it's important that you order your required quantity of handbooks as soon as possible, instead of waiting until the need arises. An early order permits the Air Force to print enough books to cover your requirements. If you delay your order, sufficient copies may not have been originally printed, so it may take a long time to fulfill your request.

To understand the Technical Order system, read *Technical Order 0J-5-2,* which explains in just a few pages how easy it is to set the system in operation. Actually, all that's required is for you to list the quantity you need on the *Publications Requirements Table (T.O. 00-3-1),* and all subsequent revisions, reissues and supplements will be automatically forwarded to you in the same quantities. (Since the preparation and distribution of handbook revision material takes time, the Air Force issues supplementary T.O.'s to reflect changes made to airplanes in service.) Check with your Base Supply Officer; he knows about the system, as it is his job to fulfill T.O. requests.

The Air Force now issues Safety of Flight Supplements to make sure you get the latest information on critical operational changes in a hurry. These supplements use the same basic T.O. number as your Flight Handbook, except for the addition of a suffix letter. Supplements covering loss of life will get to you within 48 hours after being issued; those dealing with serious damage to equipment will reach you in 6 days. If you have ordered your Flight Handbook on the Publications Requirements Table, you need do absolutely nothing to get these supplements—they'll come to you automatically.

Any comments you have regarding this handbook, suggestions for future books, or questions on any phase of the Flight Handbook program are invited and should be addressed to the Wright Air Development Center, Wright-Patterson Air Force Base, Dayton, Ohio, Attn: WCSOF.

This handbook is divided into nine sections, an appendix, and an index as follows:

Section I, DESCRIPTION—a detailed description of the airplane and the equipment and systems (including all emergency equipment not part of the auxiliary equipment) which are essential for flight.

Section II, NORMAL PROCEDURES—operating instructions arranged in proper sequence from the time you approach the airplane until it is parked after flight.

Section III, EMERGENCY PROCEDURES—concise procedures to be followed in meeting any emergency (except those of auxiliary equipment) that could reasonably be expected.

Section IV, DESCRIPTION AND OPERATION OF AUXILIARY EQUIPMENT—descriptions and normal and emergency operating instructions for all equipment not essential for flying the airplane, such as cockpit heating and ventilating, oxygen, lighting, armament, and miscellaneous equipment.

Section V, OPERATING LIMITATIONS—all airplanes and engine operating limitations that must be observed during operation.

Section VI, FLIGHT CHARACTERISTICS—a discussion ot flight characteristics, the advantageous as well as the dangerous, that are peculiar to the airplane as based on extensive flight tests.

Section VII, SYSTEMS OPERATION—a supplementary discussion of special characteristics and factors involved in operating some of the airplane systems under various conditions.

Section VIII, CREW DUTIES—omitted as not applicable for a single-place airplane.

Section IX, ALL-WEATHER OPERATION—supplementary procedures and operating instructions for safe and efficient operation under instrument flight and extreme weather conditions.

Appendix, OPERATING DATA—all operating data charts for efficient preflight and in-flight mission planning. Take-off and landing charts for various gross weights are also included.

Alphabetical Index—a complete listing of material in this handbook, including illustrations, arranged for ease in reference.

Figure 1-1

Description

SECTION I

AIRPLANE.

The F-51D, built by North American Aviation, is a single-place, low-wing monoplane powered by a Packard-built Rolls Royce engine. Although designed primarily as a fighter airplane, it may be equipped to carry bombs, chemical tanks, or rockets. Provision is also made to allow installation of external drop tanks to increase operating range. The airplane is equipped with six .50-caliber machine guns as standard equipment. Armor plate is installed in the cockpit area for protection of the pilot during combat.

AIRPLANE DIMENSIONS.

The following dimensions apply to the airplane:

Wing span ....37 feet
Length ....32 feet 2 inches
Height (three-point attitude)..13 feet 8 inches

AIRPLANE GROSS WEIGHT.

The normal gross weight of the airplane with no external load is approximately 9000 pounds. The gross weight can be as high as 12,300 pounds when external armament and fuel are carried.

ARMAMENT.

The airplane is armed with six forward-firing .50-caliber machine guns, three mounted in each wing, and is capable of carrying bombs, rockets, or chemical tanks as external load.

ENGINE.

The airplane is powered with a Packard-built Rolls Royce V-1650-3, -7 or -9A series engine rated at 1490 horsepower (Military Power) and has a two-stage, two-speed supercharger. The 12-cylinder, liquid-cooled engine drives a four-bladed constant-speed propeller and is equipped with an injection-type carburetor and an automatic manifold pressure regulator. An aneroid-actuated switch automatically controls the supercharger blower speed shift.

ENGINE CONTROLS.

THROTTLE.

The throttle (13, figure 1-4), located on the left longeron, is mechanically linked to the manifold pressure regulator. The throttle incorporates a gate that allows a maximum of 61 in. Hg manifold pressure up to critical altitude. When the throttle is moved past the gate, breaking the light locking wire, a manifold pressure of

Main DIFFERENCES TABLE

	F-51D	F-51H	TF-51D
ARMAMENT	THREE .50-CALIBER GUNS IN EACH WING, BOMBING AND ROCKET EQUIPMENT	THREE .50-CALIBER GUNS IN EACH WING, BOMBING, ROCKET, CHEMICAL, OR DEPTH CHARGES	NO ARMAMENT
FUEL CAPACITY US. GALLONS	INTERNAL CAPACITY 245 GAL. TWO 75 GAL DROP TANKS OR TWO 110 GAL DROP TANKS	INTERNAL CAPACITY 260 GAL. TWO 75 GAL DROP TANKS OR TWO 110 GAL DROP TANKS OR TWO 165 GAL DROP TANKS	180 GAL — RIGHT AND LEFT WING TANKS ONLY
RADIO AND ELECTRONICS	AN/ARC-3 COMMAND SET, SCR-695A IDENTIFICATION, BC-453B RANGE RECEIVER, AN/ARA-8 HOMING ADAPTER	AN/ARC-3 COMMAND SET, SCR-695 IDENTIFICATION, BC-453B RANGE RECEIVER, AN/ARN-7 RADIO COMPASS, AN/APS-13 TAIL WARNING RADAR	AN/ARN-6 RADIO COMPASS, AN/ARC-3 COMMAND SET, BC-453B RANGE RECEIVER, R-122/ARN-12 MARKER BEACON AND INTERPHONE
CANOPY	MANUALLY OPERATED, SLIDING TYPE	MANUALLY OPERATED, SLIDING TYPE	ELECTRICALLY OR MANUALLY OPERATED, SLIDING TYPE
ARMOR	FIRE WALL AND BACK OF SEAT ARMOR, WINDSHIELD	FIRE WALL AND BACK OF SEAT ARMOR, WINDSHIELD	FIRE WALL AND ARMOR GLASS, WINDSHIELD
ANTI-G	ANTI-G SUIT CONNECTION	ANTI-G SUIT CONNECTION	NO ANTI-G SUIT CONNECTION
ENGINE	V-1650-3, -7, OR -9A	V-1650-9	V-1650-3 OR -7
FIRE-FIGHTING EQUIPMENT	NONE	NONE	FIRE EXTINGUISHER IN FRONT COCKPIT
WHEEL BRAKES	INTERNAL SHOE	SPOT BRAKES	INTERNAL SHOE
COCKPIT ARRANGEMENT	SINGLE COCKPIT	SINGLE COCKPIT	FRONT AND REAR COCKPIT
GUN SIGHT	TYPE K-14A, K-14B, OR N-9	TYPE K-14A OR K-14B	NONE

109-00-1532

Figure 1-2

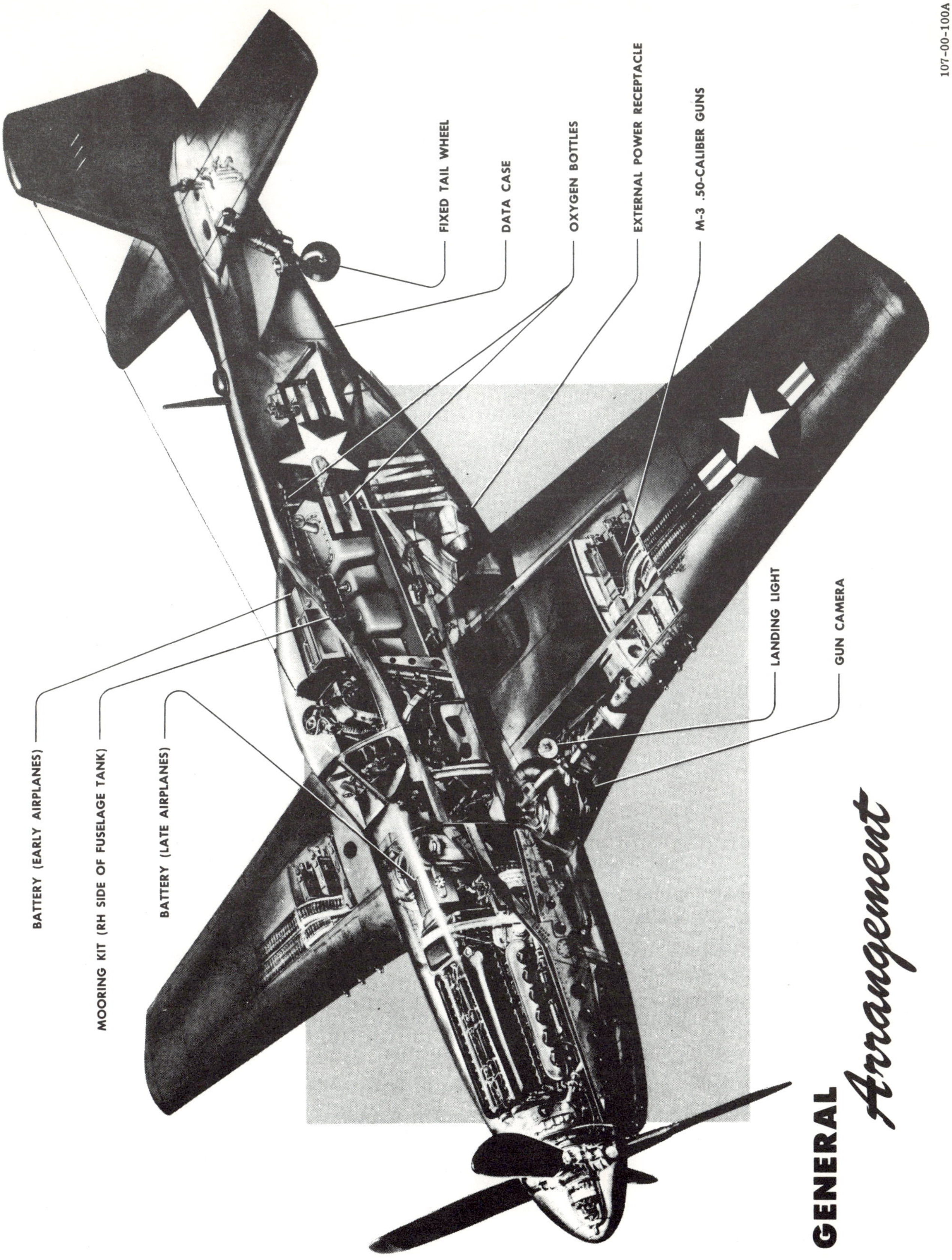
107-00-100A
FIXED TAIL WHEEL
DATA CASE
OXYGEN BOTTLES
EXTERNAL POWER RECEPTACLE
M-3 .50-CALIBER GUNS
LANDING LIGHT
GUN CAMERA
BATTERY (EARLY AIRPLANES)
MOORING KIT (RH SIDE OF FUSELAGE TANK)
BATTERY (LATE AIRPLANES)
GENERAL
Arrangement

Figure 1-3

Cockpit LEFT SIDE

1. Side Air Outlet Control Knob
2. Signal Pistol Storage Case
3. Signal Pistol Mount
4. Armrest
5. Cockpit Light
6. Coolant Radiator Air Control Switch
7. Oil Radiator Air Control Switch
8. Landing Light Switch
9. Fluorescent Light Rheostat
10. Mixture Control
11. Propeller and Mixture Control Friction Lock
12. Propeller Control
13. Throttle
14. Microphone Button
15. Bomb-Tank Salvo Levers
16. Landing Gear Handle
17. Throttle Friction Lock
18. Aileron Trim Tab Control Knob
19. Elevator Trim Tab Control Wheel
20. Rudder Trim Tab Control Knob
21. Shoulder-harness Lock Handle
22. Carburetor Ram-air Control Lever
23. Carburetor Hot-air Control Lever
24. Anti-G Suit Connector
25. Wing Flap Handle
26. Map Case
27. Drop Message Bag Holder

109-00-1476

Figure 1-4

Cockpit ... RIGHT SIDE

1. Oxygen Regulator
2. Canopy Emergency Release Handle
3. Circuit-breaker Reset Cover
4. Recognition Light Keying Switch (Disconnected)
5. Fluorescent Light Rheostat
6. Ammeter
7. Canopy Handcrank
8. VHF Control Box
9. DF Tone Switch
10. IFF Radar Control Panel
11. Side Air Outlet Control Knob
12. Oxygen Hose
13. Coolant Flap Emergency Release Handle
14. First-aid Kit
15. Circuit Breakers
16. Seat Adjustment Lever
17. Cockpit Light
18. VHF Volume Control Knob
19. Range Receiver Control Box
20. Right Switch Panel
21. Battery-disconnect Switch
22. Generator-disconnect Switch
23. Spare Bulb Compartment

109-00-1477

Figure 1-5

Cockpit FORWARD VIEW

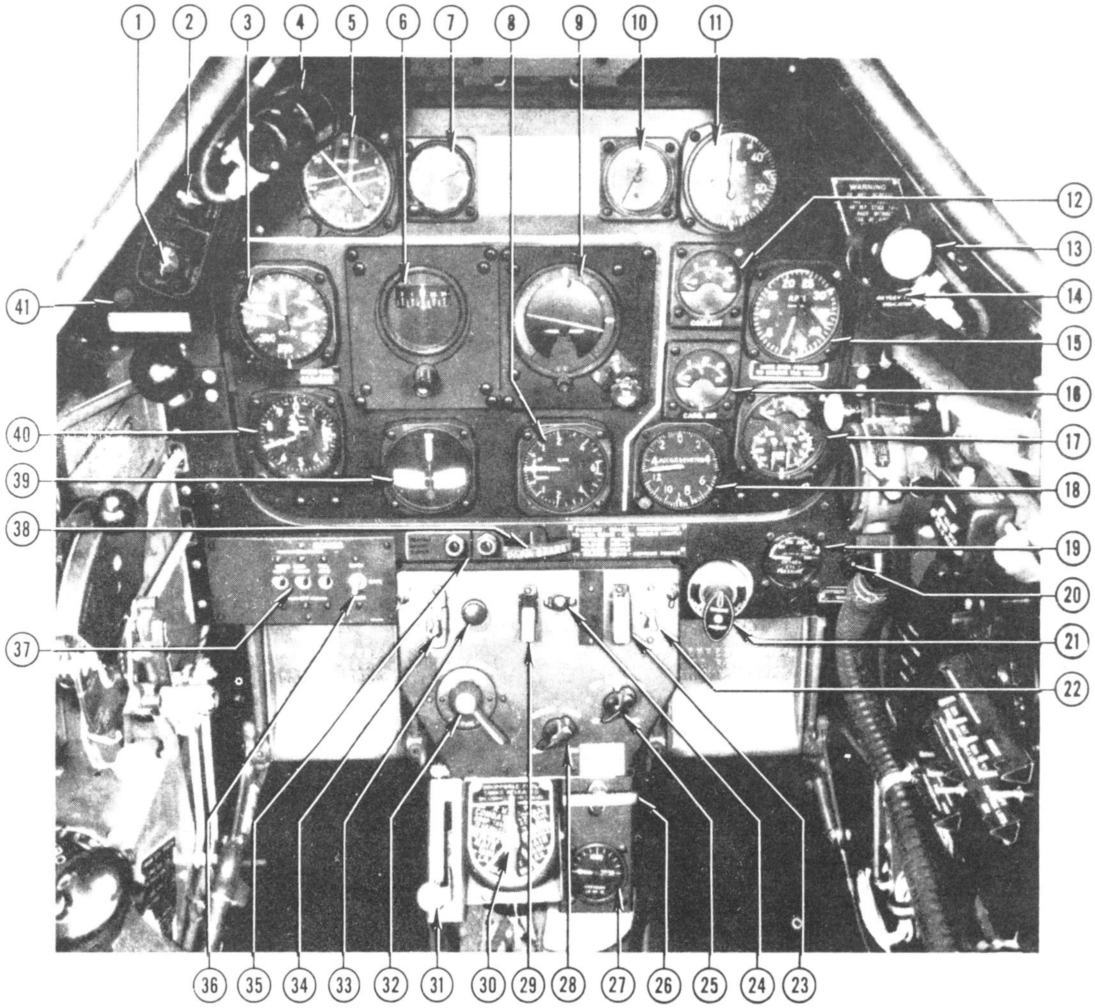

1. K-14 Gun Sight Gyro Selector Switch
2. K-14 Gun Sight Dimmer Rheostat
3. Airspeed Indicator
4. Fluorescent Light
5. Remote-reading Compass Indicator
6. Directional Gyro
7. Clock
8. Rate-of-Climb Indicator
9. Flight Indicator
10. Suction Gage
11. Manifold Pressure Gage
12. Coolant Temperature Gage
13. Fluorescent Light
14. Oxygen Flow Blinker
15. Tachometer
16. Carburetor Air Temperature Gage
17. Engine Gage Unit
18. Accelerometer (Radio Compass on Some Airplanes)
19. Oxygen Pressure Gage
20. Oxygen Low-pressure Warning Light
21. Engine Primer (Early Airplanes)
22. Oil Dilution Switch
23. Starter Switch
24. Supercharger High-blower Warning Light
25. Gun Sight Rheostat (N-9)
26. Fairing Door Emergency Release Handle
27. Hydraulic Pressure Gage
28. Cockpit Light Rheostat
29. Supercharger Control Switch
30. Fuel Tank Selector Handle
31. Fuel Shutoff Lever
32. Ignition Switch
33. Warning Horn Silencer Button
34. Fuel Booster Pump Switch
35. Landing Gear Warning Lights
36. Bomb Release Selector Switch
37. Bomb-arming Switches
38. Parking Brake Handle
39. Turn-and-Bank Indicator
40. Altimeter
41. Landing Gear Warning Signal Test Switch

109-00-1475

Figure 1-6

as much as 67 in. Hg is possible for War Emergency Power. A throttle friction lock (17, figure 1-4) on the throttle quadrant is provided to hold the throttle in any desired position. Early airplanes* have a ball-type knob with a microphone button inset into the ball. Late airplanes† have a grip-type throttle handle, which rotates to adjust range of the gun sight. A microphone button is installed at the end of the grip.

MIXTURE CONTROL.

The mixture control (10, figure 1-4), located on the aft side of the throttle quadrant, has three positions: IDLE CUTOFF, NORMAL, and RICH.

CAUTION

> The mixture control should always be in the IDLE CUTOFF position when the engine is not running, to prevent fuel from entering the carburetor.

CARBURETOR AIR.

Cold outside ram air to the carburetor enters a duct in the nose just below the propeller spinner. (See figures 1-9 and 1-10.) A door at the forward end of the duct can be closed mechanically from the cockpit to force the air to enter through a perforated side panel (and filter) on each side of the engine cowl. For cold-weather operation, these perforated side panels can be replaced with blank panels. With blank panels installed, the induction system is forced to pull warm air from the engine compartment, through a spring-loaded door, whenever the ram-air door is closed. On late airplanes,† this spring-loaded door is also mechanically operated from the cockpit to permit warm air to enter as desired by the pilot. If at any time the ram-air duct becomes clogged with ice, warm air from the engine compartment will automatically be admitted.

CARBURETOR RAM-AIR CONTROL LEVER.

The carburetor ram-air control lever (22, figure 1-4), located on the left console aft of the rudder trim control knob, opens and closes the ram-air door in the carburetor air duct. The lever has two positions, RAM AIR and UNRAMMED FILTERED AIR. When the ram-air control lever is moved to the RAM AIR position, the ram-air door opens and permits ram air to enter the carburetor. Moving the ram-air control lever to UNRAMMED FILTERED AIR closes the ram-air door and the air passes through the filter to the carburetor.

Figure 1-7

CARBURETOR HOT-AIR CONTROL LEVER.†

In addition to the ram-air control lever, a carburetor hot-air control lever (23, figure 1-4) is installed on late airplanes.† The hot-air control lever, located adjacent to the carburetor ram-air control lever, has two positions, NORMAL and HOT AIR. When the hot-air control lever is positioned at NORMAL, cold air enters the carburetor. Moving the hot-air control lever to HOT AIR opens the hot-air door and permits warm air to enter the carburetor air duct if the ram-air control lever is positioned at UNRAMMED FILTERED AIR.

WARNING

> Because of adverse leaning effect, carburetor hot air should not be used above 12,000 feet altitude. The heat affects the altitude compensator of the carburetor.

***F-51D-5-NA Through F-51D-30-NA and F-51D-5-NT Airplanes**
†F-51D-20-NT and subsequent airplanes

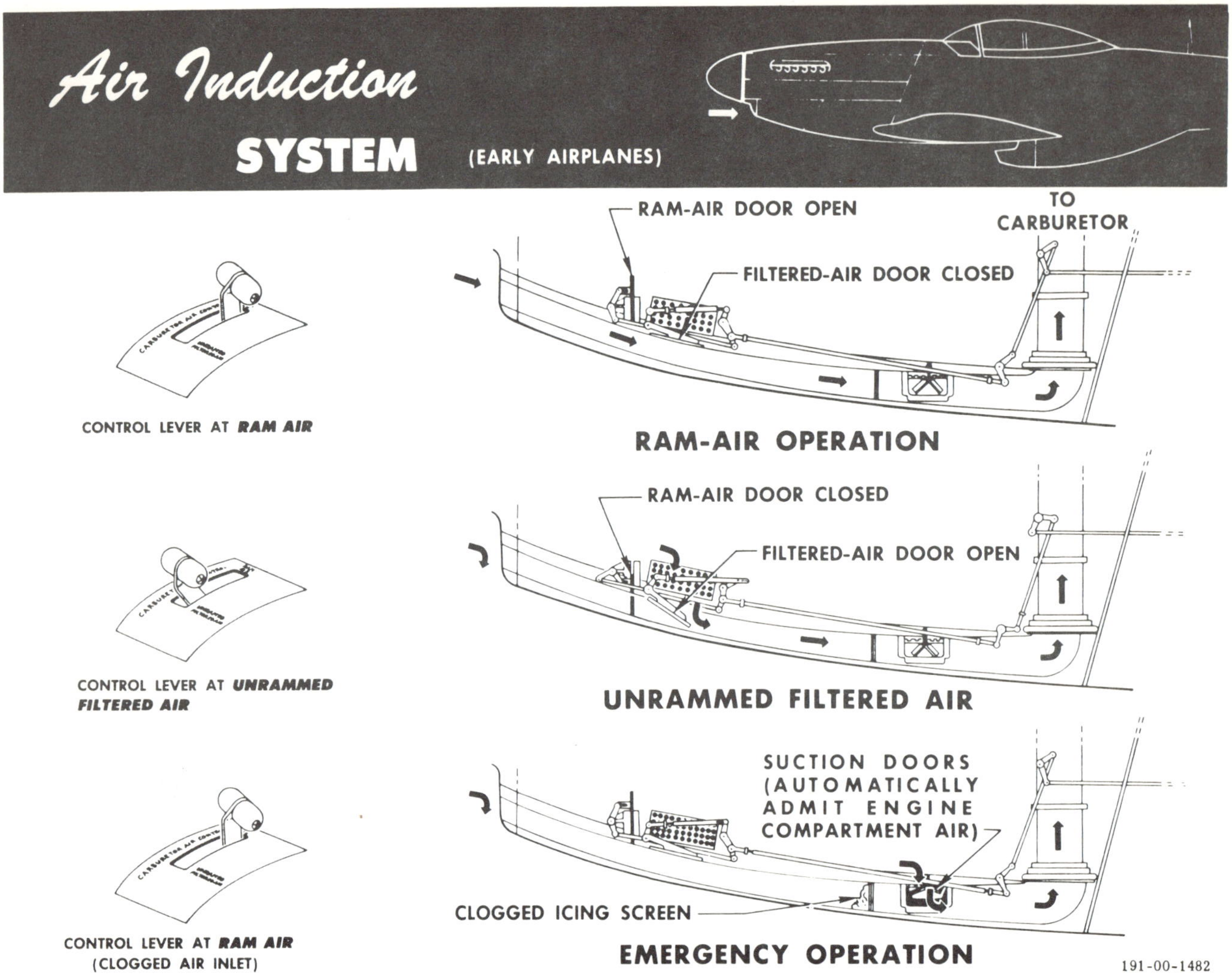

Figure 1-9

COOLING SYSTEMS.

There are two complete and separate cooling systems used in this airplane. One system, the engine cooling system, is used to cool the engine. The other, the aftercooling system, cools the supercharged fuel-air mixture. Coolant from each system passes through the respective portion of the dual radiator, located in the aft portion of the air scoop on the underside of the fuselage. The radiator is actually two radiators constructed as a single unit with separate cores. An electrically controlled flap-type door is used to control the airflow through the radiator. In case of actuator failure, an emergency handle opens the coolant flap to lower the temperature. The coolant solution consists of water and ethylene glycol in varying percentages depending on outside operating temperatures.

ENGINE COOLING SYSTEM. An engine-driven pump pressurizes the engine cooling system (figure 1-11), which has a capacity of 16.7 gallons including 5.2 gallons in the coolant header tank. The system may be filled at the coolant header tank, which is accessible through the dzus-fastened door at the forward end of the engine upper left cowling. (See figure 1-25.)

AFTERCOOLING SYSTEM. The aftercooling system (figure 1-11) is a low-pressure type and has a system capacity of 4.8 gallons. This capacity includes the aftercooling header tank, which contains ½ gallon. (See figure 1-25.) Coolant is forced by an engine-driven pump through the radiator to the supercharger cooling jackets, and from there returns to the aftercooler unit. Cooling the fuel-air mixture before it enters the combustion chambers of the engine minimizes the possibility of detonation.

COOLANT RADIATOR AIR CONTROL SWITCH.

Airflow through the dual radiator is controlled by an electric actuator which is mechanically connected to

Air Induction SYSTEM

(LATE AIRPLANES)

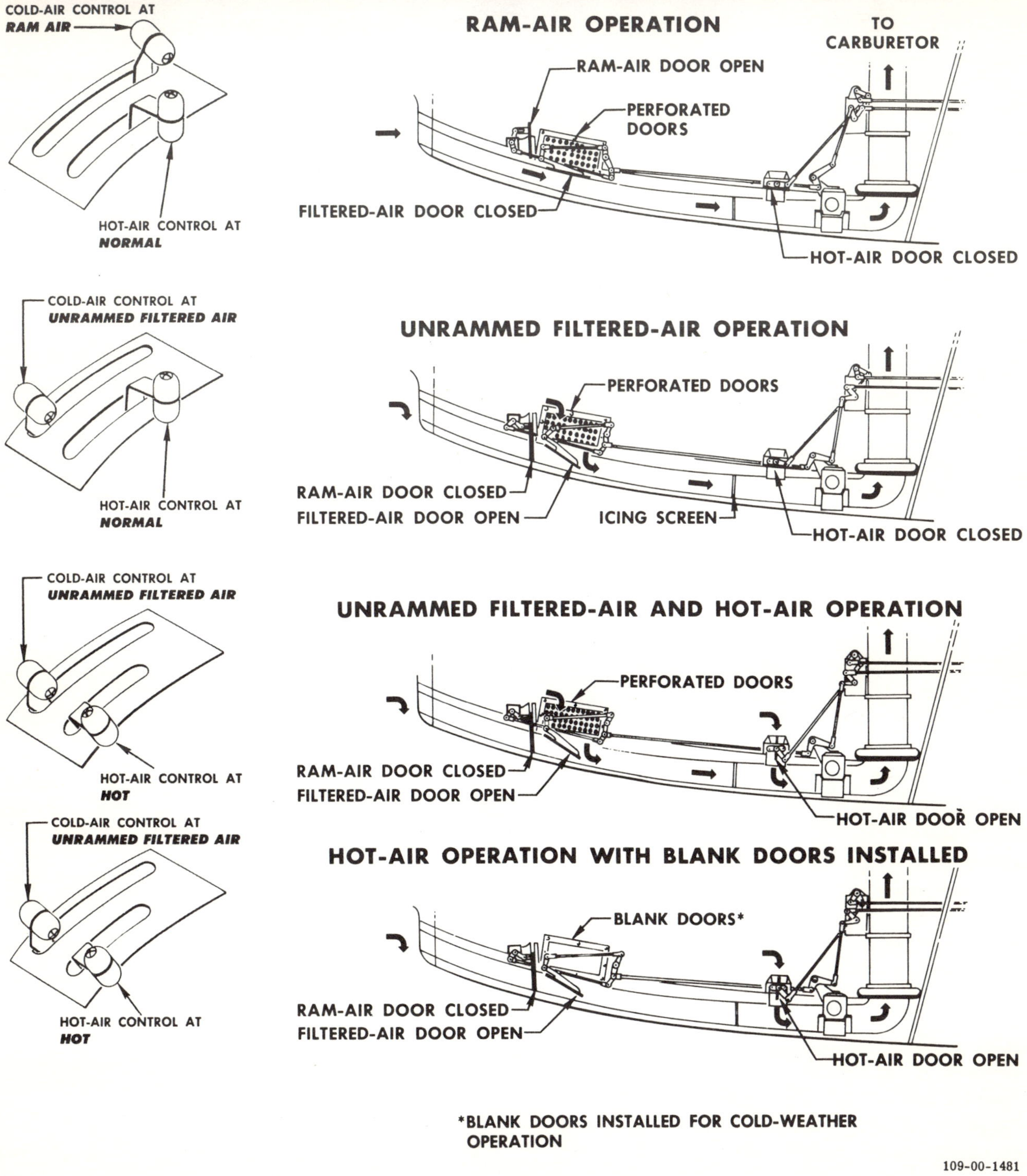

Figure 1-10

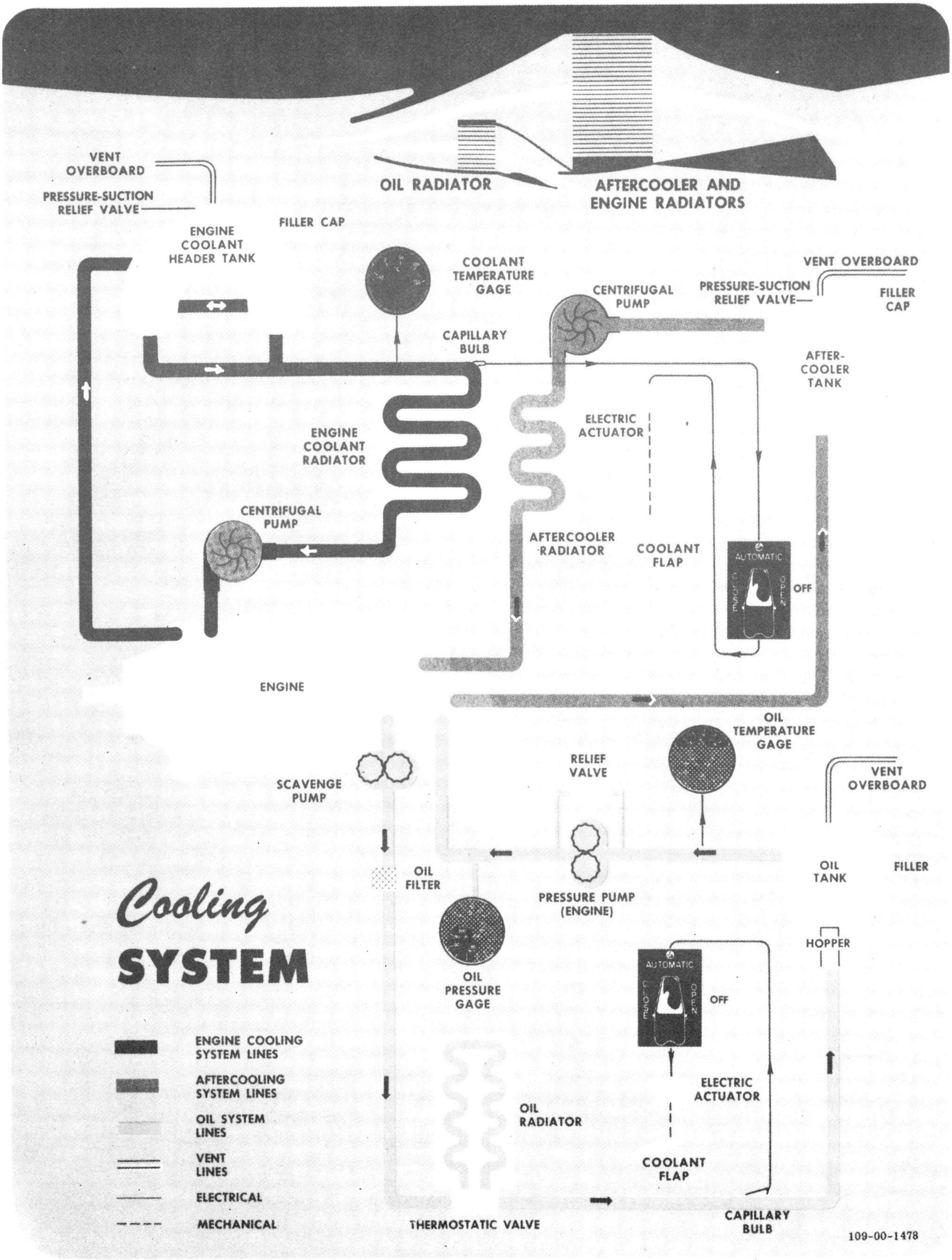
VENT OVERBOARD
PRESSURE-SUCTION RELIEF VALVE
OIL RADIATOR
AFTERCOOLER AND ENGINE RADIATORS
FILLER CAP
ENGINE COOLANT HEADER TANK
COOLANT TEMPERATURE GAGE
VENT OVERBOARD
CENTRIFUGAL PUMP
PRESSURE-SUCTION RELIEF VALVE
FILLER CAP
CAPILLARY BULB
AFTER-COOLER TANK
ELECTRIC ACTUATOR
ENGINE COOLANT RADIATOR
CENTRIFUGAL PUMP
AFTERCOOLER RADIATOR
COOLANT FLAP
AUTOMATIC
CLOSE
OPEN
OFF
ENGINE
OIL TEMPERATURE GAGE
RELIEF VALVE
SCAVENGE PUMP
VENT OVERBOARD
OIL FILTER
PRESSURE PUMP (ENGINE)
OIL TANK
FILLER
Cooling SYSTEM
HOPPER
OIL PRESSURE GAGE
AUTOMATIC
CLOSE
OPEN
OFF
ENGINE COOLING SYSTEM LINES
AFTERCOOLING SYSTEM LINES
OIL SYSTEM LINES
VENT LINES
ELECTRICAL
MECHANICAL
ELECTRIC ACTUATOR
OIL RADIATOR
COOLANT FLAP
THERMOSTATIC VALVE
CAPILLARY BULB
109-00-1478

Figure 1-11

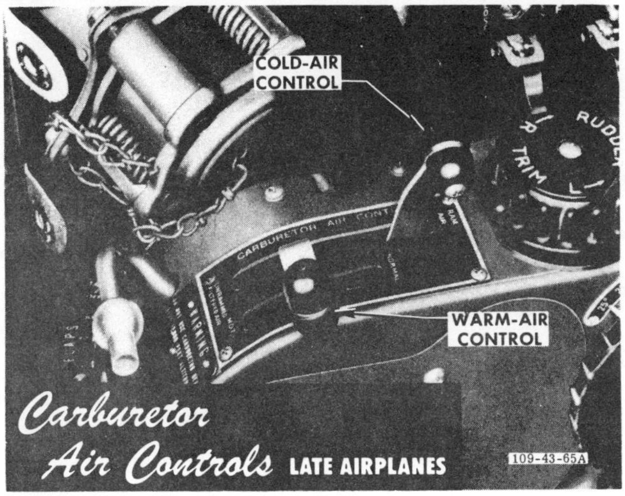

Figure 1-12

the coolant flap. The operation of the actuator is controlled by a switch (6, figure 1-4; figure 1-13) located on the left side of the cockpit just aft of the throttle quadrant. The switch has four positions: AUTOMATIC, OPEN, CLOSE, and OFF. The AUTOMATIC position is used for normal operation; the switch is held in this position by a spring-loaded guard. With the switch in this position, the temperature of the coolant governs the amount the coolant flap will be opened or closed. From the spring-loaded OPEN or CLOSE position, the switch returns to OFF when released. The OPEN and CLOSE positions permit opening and closing the coolant flap as desired by the pilot during ground operation or if manual adjustment is necessary during flight. For all ground operation, the switch should be held at OPEN until coolant flap is fully opened, then released to OFF.

COOLANT FLAP EMERGENCY RELEASE HANDLE.

A manual coolant flap emergency release handle (13, figure 1-5) is located on the floor of the cockpit, to the right of the seat. In case of actuator failure, a quick pull of this handle mechanically extends the coolant flap an additional 5½ inches by increasing the length of the linkage to the coolant flap. If the coolant flap is completely closed, the flap will open to a minimum of 7 inches. After the emergency release has been pulled, there is no means of resetting it in flight.

SUPERCHARGER.

The engine-driven, two-speed, two-stage supercharger is of the centrifugal type automatically controlled by an aneroid-operated switch vented to the carburetor intake pressure. The aneroid switch changes the blower speed from low to high at the altitude for best performance at Military Power. The switch is calibrated to shift the supercharger to high blower on the V-1650-3 and V-1650-9A engines between 20,800 and 24,800 feet and on the V-1650-7 engine between 15,700 and 19,700 feet. To prevent excessively frequent blower speed changes resulting from small speed or altitude changes near shift altitude, the aneroid switch is constructed so that the shift downward from high to low speed occurs approximately 1500 feet below the upward shift point during a normal descent. However, during a dive or rapid descent, the shift downward may occur at or above the upward shift point because of the increase in ram-air pressure at the carburetor intake due to the higher airspeed.

Note

In flight, the blower may shift at altitudes other than specified in the preceding paragraph for the particular engine. This condition is normal, since the blower shift aneroid is referenced to carburetor entrance air pressure, which increases with indicated airspeed. Differences in airplane altitude at the time of blower shift are due to the ram-air variations in climb, level flight, and descent.

For minimum fuel consumption on long-range cruising operations, it is advantageous to remain in low blower speed above the altitude of shift. The ranges shown on the charts in Appendix I are possible only when proper supercharger speed is used. In case of blower shift aneroid failure, the supercharger automatically returns to low speed.

SUPERCHARGER CONTROL SWITCH.

The supercharger control switch (29, figure 1-6; figures 1-14 and 1-15), mounted on the front switch panel, has three positions; LOW, AUTO, and HIGH. A spring-loaded guard holds the switch in the AUTO position. When the switch is at the AUTO position, the supercharger is controlled by an electrical aneroid-type pressure switch vented to the carburetor intake pressure. For all normal operation, the switch should be at AUTO. The LOW position is used to open the circuit to the supercharger solenoid for low-blower operation in the event the aneroid switch fails. The HIGH position permits shifting to high blower below the preset shift altitude and to operate high blower for ground checks.

SUPERCHARGER HIGH-BLOWER WARNING LIGHT.

An amber light (24, figure 1-6; figures 1-14 and 1-15) is provided on the front switch panel beside the supercharger control switch to indicate when the supercharger is in high blower. On late airplanes, the light is of the push-to-test type.

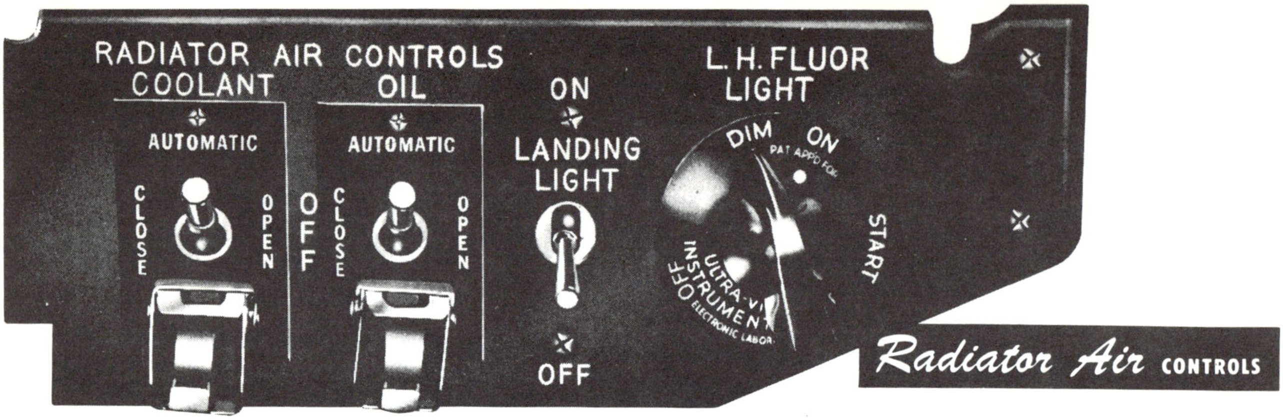

109-54-133

Figure 1-13

IGNITION SYSTEM.

Two engine-driven high-tension magnetos, mounted on the engine, supply spark for combustion and are grounded when the ignition system is inoperative. Both magnetos have booster coil connections, but only the one on the right magneto is used. The booster coil intensifies the spark of the right magneto to aid in starting.

IGNITION SWITCH.

The ignition switch (32, figure 1-6; figure 1-14) is located on the front switch panel and has four positions: OFF, R, L, and BOTH.

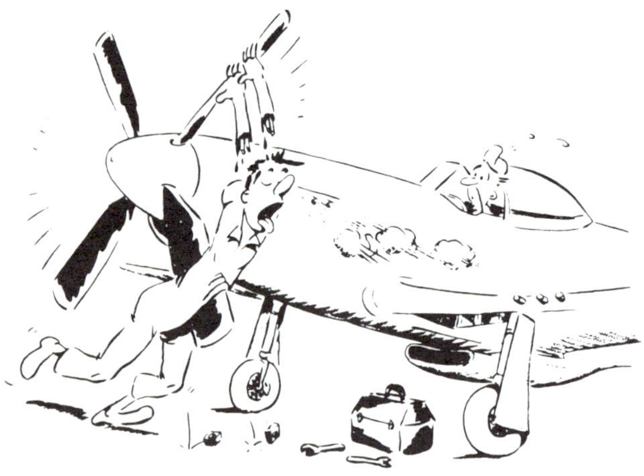

Caution

To prevent accidental engine start, be sure ignition switch is moved to OFF after IDLE CUTOFF position of mixture control is used for stopping engine.

PRIMER SYSTEM.

The electrically energized priming system gets its fuel from the engine-driven fuel pump and consists of a primer, a solenoid valve, a momentary-on switch, and connections to the induction manifold. On early airplanes incorporating a hand-operated primer, the primer pump withdraws fuel from the fuel strainer on the fire wall and directs it to the induction manifold.

PRIMER SWITCH.

Some airplanes incorporate an electric primer switch. The primer switch (figure 1-14; figure 1-15) is located on the front switch panel and has two positions, OFF and momentary ON. When the primer switch is held ON, the solenoid valve mounted on the carburetor permits fuel to pass to the primer lines and into the induction manifold. Usually 3 or 4 seconds is sufficient to prime a cold engine. The engine should be primed only when it is turning over.

HAND-OPERATED PRIMER.

Early airplanes have a hand-operated primer (21, figure 1-6), located on the lower right side of the instrument panel. To unlock the primer, the handle is depressed and rotated. For priming a cold engine, the primer should be pumped a maximum of three or four strokes. Priming should be accomplished only when the engine is turning over.

Note

Be sure primer is returned to locked position after priming.

STARTER SYSTEM.

The starter system consists of an electric direct-cranking starter, a starter switch, and a booster coil. To aid the magnetos when rpm is low during cranking, a booster coil intensifies the spark of the right magneto, which fires the intake spark plug of each cylinder.

STARTER SWITCH.

The starter switch (23, figure 1-6; figures 1-14 and 1-15), located on the front switch panel, has an ON position and a spring-loaded guarded OFF position. Holding the switch at ON energizes both the starter and the booster coil.

AUTOMATIC MANIFOLD PRESSURE REGULATORS.

The V-1650-3 and early -7 engines have a Packard manifold pressure regulator which is mechanically operated by pressure and temperature through an aneroid unit that mechanically operates the throttle linkage. This regulator maintains a constant manifold pressure within ±1 inch between 42 and 61 in. Hg for all altitudes below the critical altitude of the engine. Below 42 in. Hg, the manifold pressure regulator cannot be expected to maintain a constant manifold pressure. On the V-1650-9A and late -7 engines, a Simmond's automatic engine control unit is used. This control unit, using engine oil for operation, automatically maintains a constant manifold pressure at all power settings between 25 and 67 in. Hg (±1 inch) up to the critical altitude of the engine. If at any time the operating oil to the control unit should fail, the unit becomes fully manual over the entire range of manifold pressures available. The maximum pressure available at this time is approximately 52 in. Hg. To prevent a runaway engine during starting, a manual override is mechanically linked to the control unit from the throttle to manually close the butterfly valve. Stopping and starting procedures must be strictly followed to prevent a runaway engine during starting. The advantage of the automatic control unit in constantly maintaining a selected manifold pressure, more than compensates for the difficulty of carburetor ice detection and strict stopping and starting procedures.

ENGINE INDICATORS.

Standard engine instruments are provided in the airplane. The oil pressure, fuel pressure, and manifold pressure gages indicate pressure readings directly from the engine. The tachometer is self-generated. Power from the airplane electrical system is required to operate the oil temperature, coolant temperature, and carburetor temperature indicators

PROPELLER.

The airplane is equipped with an 11-foot 2-inch diameter, four-bladed, constant-speed, Hamilton Standard Hydromatic propeller of the nonfeathering type. A propeller governor mechanically controlled from the cockpit admits engine oil to the propeller dome for pitch changes necessary to maintain a constant engine speed. Engine oil pressure is used to aid the centrifugal twisting moment of the blades toward low pitch (increase rpm), and boosted engine oil pressure through the governor moves the blades toward high pitch (decrease rpm).

PROPELLER CONTROL.

A propeller control (12, figure 1-4), located on the throttle quadrant, is mechanically linked to the governor. The control setting determines the engine rpm, which is maintained constant by the propeller governor. The propeller control may be positioned at INCREASE or DECREASE or to any intermediate position.

OIL SYSTEM.

Oil for engine lubrication is supplied from a 12.5 US. gallon tank. (See figure 1-11.) Lubrication is accomplished by a pressure system with a dry sump and scavenge pump return. Oil flows by gravity from the

109-00-1472

Figure 1-14

Engine Control Panel
AIRPLANES WITH ZERO RAIL ROCKET INSTALLATION

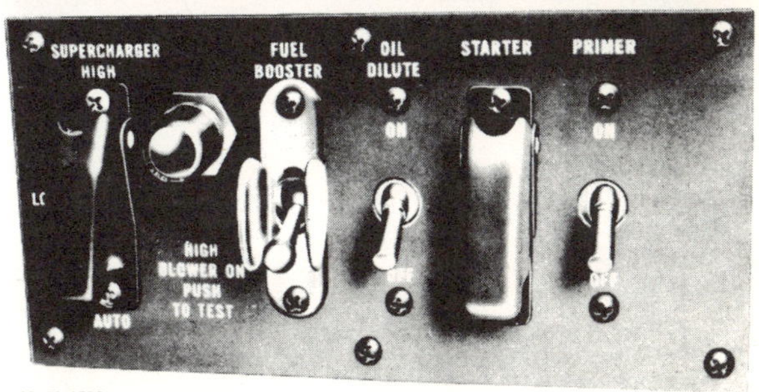

109-00-1526

Figure 1-15

tank to the engine pressure pump, which forces it through the engine. The temperature of the oil is regulated automatically under normal conditions. The oil either flows through the oil radiator when cooling is necessary or flows directly back to the oil tank hopper unit. If the automatic temperature control fails, the oil radiator air outlet flap may be operated electrically. See figure 1-25 for oil grade and specification. A dip-stick type gage adjacent to the filler neck is used to determine the oil quantity. An oil dilution system is provided to facilitate cold-weather starting.

OIL SYSTEM CONTROLS.

OIL RADIATOR AIR CONTROL SWITCH.

A switch (7, figure 1-4; figure 1-13) to control the electric actuator of the oil radiator air outlet flap is located on the radiator control panel. The switch has four positions: OFF, CLOSE, AUTOMATIC, and OPEN. For normal operation, the switch should be left in the AUTOMATIC position. When switch is at AUTOMATIC, a thermostat automatically starts and stops the actuator to move the oil radiator air outlet flap, depending on oil temperature. However, should the automatic feature fail, an abnormal reading will show on the oil temperature gage and the switch may be moved to either the OPEN or CLOSE position as needed. Limit switches stop the actuator when full open or closed position is reached.

Fuel Quantity DATA

	NO.	USABLE FUEL IN LEVEL FLIGHT (EACH)	FULLY SERVICED (EACH)	EXPANSION SPACE (EACH)	TOTAL VOLUME (EACH)
LH MAIN	1	90.4	92.7	5.0	97.7
RH MAIN	1	90.0	92.1	4.5	96.6
FUSELAGE	1	65	65.5	26.0*	91.5
DROP TANKS	2	75.0	75.0	2.3	77.3
	OR				
	2	110.0	110.0	3.3	113.0

TOTAL USABLE INTERNAL FUEL—245.4 GAL
TOTAL USABLE FUEL WITH 75 GAL DROP TANKS—395.4 GAL
TOTAL USABLE FUEL WITH 110 GAL DROP TANKS—455.4 GAL

*FUSELAGE TANK RESTRICTED TO 65 GALLONS BECAUSE OF ADVERSE CG CONDITION

109-00-1473

Figure 1-16

OIL DILUTION SWITCH.

The oil dilution switch (22, figure 1-6; figure 1-15), located below the main instrument panel, has ON and OFF positions. Engine fuel is allowed to enter the oil system at the oil drain when the switch is in the ON position, lowering the oil viscosity for cold-weather starting. Oil should not be diluted in excess of 10 percent. Refer to "Stopping Engine" in Section IX.

OIL SYSTEM INDICATOR.

A combination oil temperature, oil pressure, and fuel pressure gage (17, figure 1-6) is installed on the right side of the main instrument panel.

FUEL SYSTEM.

The fuel system (figure 1-17) consists of two self-sealing wing tanks of 90 US. gallons (usable fuel) each and one fuselage self-sealing tank of 85 US. gallons. Filling the fuselage tank to the 85-gallon capacity may cause an aft CG condition beyond limits; therefore, the fuselage fuel tank is placarded for 65 US. gallons maximum and should never be filled above this quantity. See figure 1-16 for fuel quantity data and figure 1-25 for fuel grade and specification. Each of the main fuel tanks has its own gravity-fed submerged-type booster pump receiving power from the electrical system of the airplane. These pumps supplement the engine-driven fuel pump and will handle the fuel needs of the engine at all altitudes if the engine-driven fuel pump fails. If the booster pumps fail, the engine-driven fuel pump will supply fuel only up to approximately 8500 feet. To prevent wing heaviness, fuel should be used alternately from the wing tanks. There are provisions beneath each wing for carrying either two 75-gallon or two 110-gallon drop tanks.

CAUTION

If installation of 110-gallon drop tanks is necessary to accomplish a particular mission, maneuvers are limited to those that are absolutely necessary to conduct normal flight, because of possible structural failure.

The drop tanks have no booster pumps, but fuel is forced from them by a controlled pressure of 5 psi from the exhaust side of the vacuum pump. This pressurization will permit satisfactory flow of fuel from the drop tanks at all altitudes. If the pressure to the drop tanks fails, the engine-driven fuel pump is capable of pulling fuel from the drop tanks up to approximately 10,000 feet.

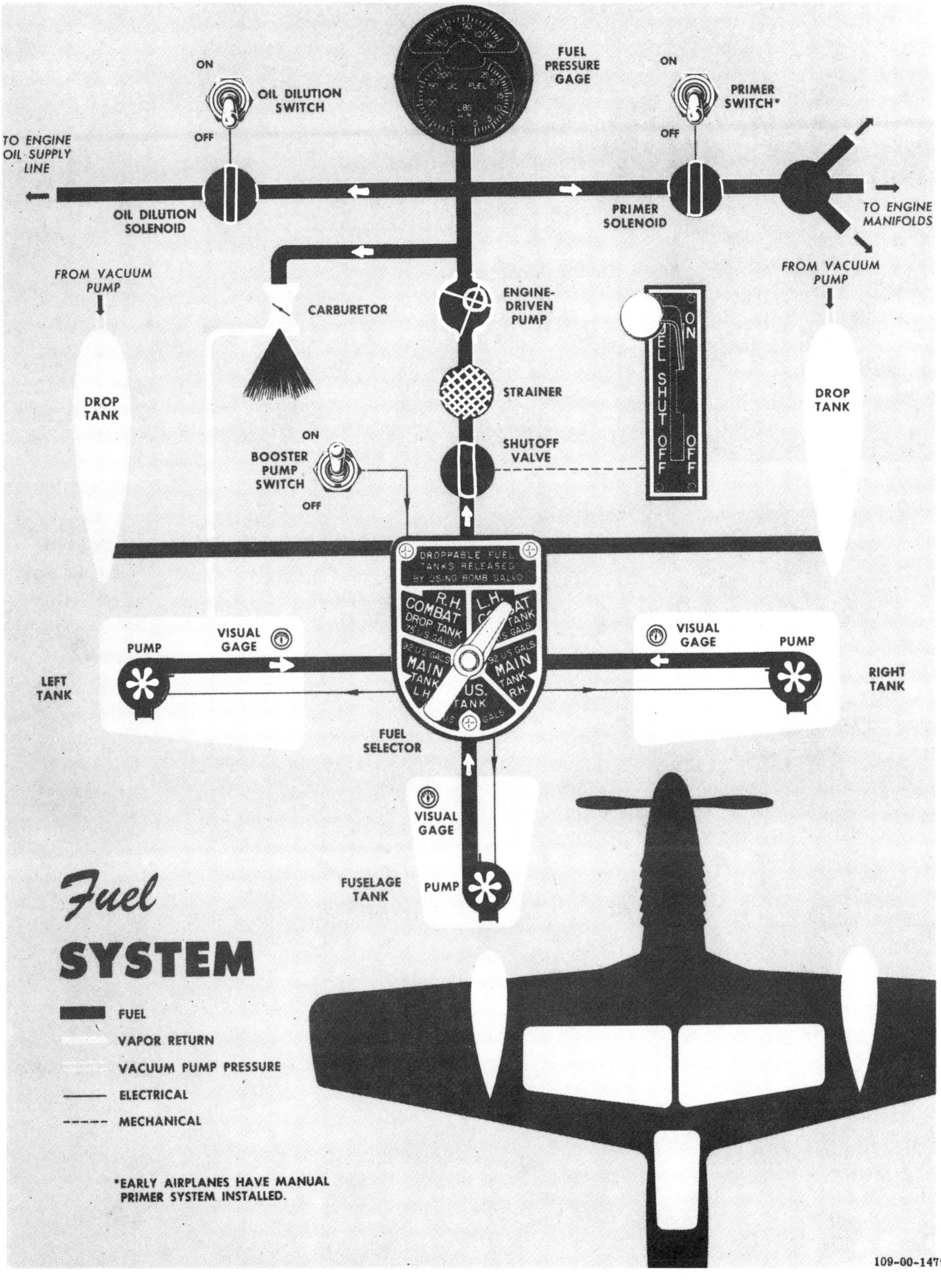
FUEL PRESSURE GAGE
ON
OIL DILUTION SWITCH
OFF
TO ENGINE OIL SUPPLY LINE
OIL DILUTION SOLENOID
ON
PRIMER SWITCH*
OFF
PRIMER SOLENOID
TO ENGINE MANIFOLDS
FROM VACUUM PUMP
CARBURETOR
ENGINE-DRIVEN PUMP
STRAINER
SHUTOFF VALVE
FUEL SHUT OFF
ON
OFF
FROM VACUUM PUMP
DROP TANK
DROP TANK
ON
BOOSTER PUMP SWITCH
OFF
DROPPABLE FUEL TANKS RELEASED BY USING BOMB SALVO
R.H. COMBAT DROP TANK 75 U.S. GALS.
L.H. COMBAT TANK
92 U.S. GALS. MAIN TANK L.H.
92 U.S. GALS. MAIN TANK R.H.
U.S. TANK
VISUAL GAGE
PUMP
LEFT TANK
VISUAL GAGE
PUMP
RIGHT TANK
FUEL SELECTOR
VISUAL GAGE
FUSELAGE TANK
PUMP
Fuel SYSTEM
FUEL
VAPOR RETURN
VACUUM PUMP PRESSURE
ELECTRICAL
MECHANICAL
*EARLY AIRPLANES HAVE MANUAL PRIMER SYSTEM INSTALLED.
109-00-1479

Figure 1-17

Fuel SHUTOFF AND SELECTOR VALVES

Figure 1-18

FUEL SYSTEM CONTROLS AND INDICATORS.

FUEL SHUTOFF LEVER.

A manually operated fuel shutoff lever (31, figure 1-6; figure 1-18) is located below and to the left of the front switch panel. The switch is mechanically linked to the fuel shutoff valve in the left wheel well. The fuel shutoff valve shuts off the fuel from all tanks to the engine-driven fuel pump.

FUEL TANK SELECTOR HANDLE.

The fuel tank selector handle (30, figure 1-6; figure 1-18) is located in the center, below the front switch panel. The following positions are marked on the mounting plate: FUS. TANK, MAIN TANK L. H., MAIN TANK R. H., R. H. COMBAT DROP TANK, and L. H. COMBAT DROP TANK. The handle is mechanically connected to the fuel selector valve in the left wheel well. A switch incorporated in the selector handle assembly is connected in series with the booster pump switch and will start the booster pump in the tank selected, provided the fuel booster pump switch is ON. Booster pumps in fuel tanks not actually supplying fuel are automatically shut off.

FUEL BOOSTER PUMP SWITCH.

A fuel booster pump switch (34, figure 1-6; figures 1-14 and 1-15), located below the instrument panel, is wired in series with the fuel selector electrical system. The switch has ON and OFF positions and must be in the ON position before the selected tank booster pump will operate. On early airplanes,* the booster pump switch has NORMAL, OFF, and EMERG. positions. The booster pump switch must be in the NORMAL position to allow pump operation on these airplanes. In the event of engine-driven pump failure, placing the switch in the EMERG. position causes the booster pumps to increase their speed, thereby increasing pressure and ensuring fuel flow to the carburetor.

BOMB OR DROP TANK SALVO LEVERS.

The drop tanks can be released whenever desired by pulling the two bomb-tank salvo levers (15, figure 1-4) located on the left side of the cockpit, below the throttle quadrant. The bomb-tank salvo levers provide a selective mechanical release of the drop tanks or bombs independent of the electrical bomb release system.

FUEL QUANTITY GAGES.

Direct-reading fuel quantity gages are installed for the internal tanks. The gages for the right and left wing tanks are located on each side of the seat, in view of the pilot. The fuselage tank gage is located behind and to the left of the pilot's seat. No fuel quantity gages are installed for the drop tanks.

ELECTRICAL POWER SUPPLY SYSTEM.

The airplane electrical system (figure 1-19) receives power from a 28-volt, 100-ampere, engine-driven generator. A 24-volt storage battery supplies current when the generator output drops below 26.5 volts. An external power receptacle is located on the right side of the fuselage, just behind the cockpit. An external power source (C-13A or equivalent) instead of battery power should be used on the ground to start the engine or to operate the electrical system when the engine is shut down. An adapter for connecting the British-type external power supply is stowed adjacent to the external power plug. All of the electrical circuits are protected by circuit-breaker switches. An inverter supplies 400-cycle, 26-volt alternating current for operation of the remote-reading compass. The inverter receives its power directly from the battery whenever the battery-disconnect switch is ON.

SWITCH PANEL AND CIRCUIT BREAKERS.

The main switches are located on the right switch panel. Location of main electrical switches is shown in figures 1-5 and 1-6. Circuit breakers on the right switch panel protect the electrical system. A single "bump" plate (3, figure 1-5) permits the pilot to reset all the circuit breakers at one time. If a circuit breaker repeatedly pops out, the master switch for that circuit should be turned OFF.

ELECTRICALLY OPERATED EQUIPMENT.

See figure 1-19.

***F-51D-5-NA and F-51D-5-NT Airplanes**

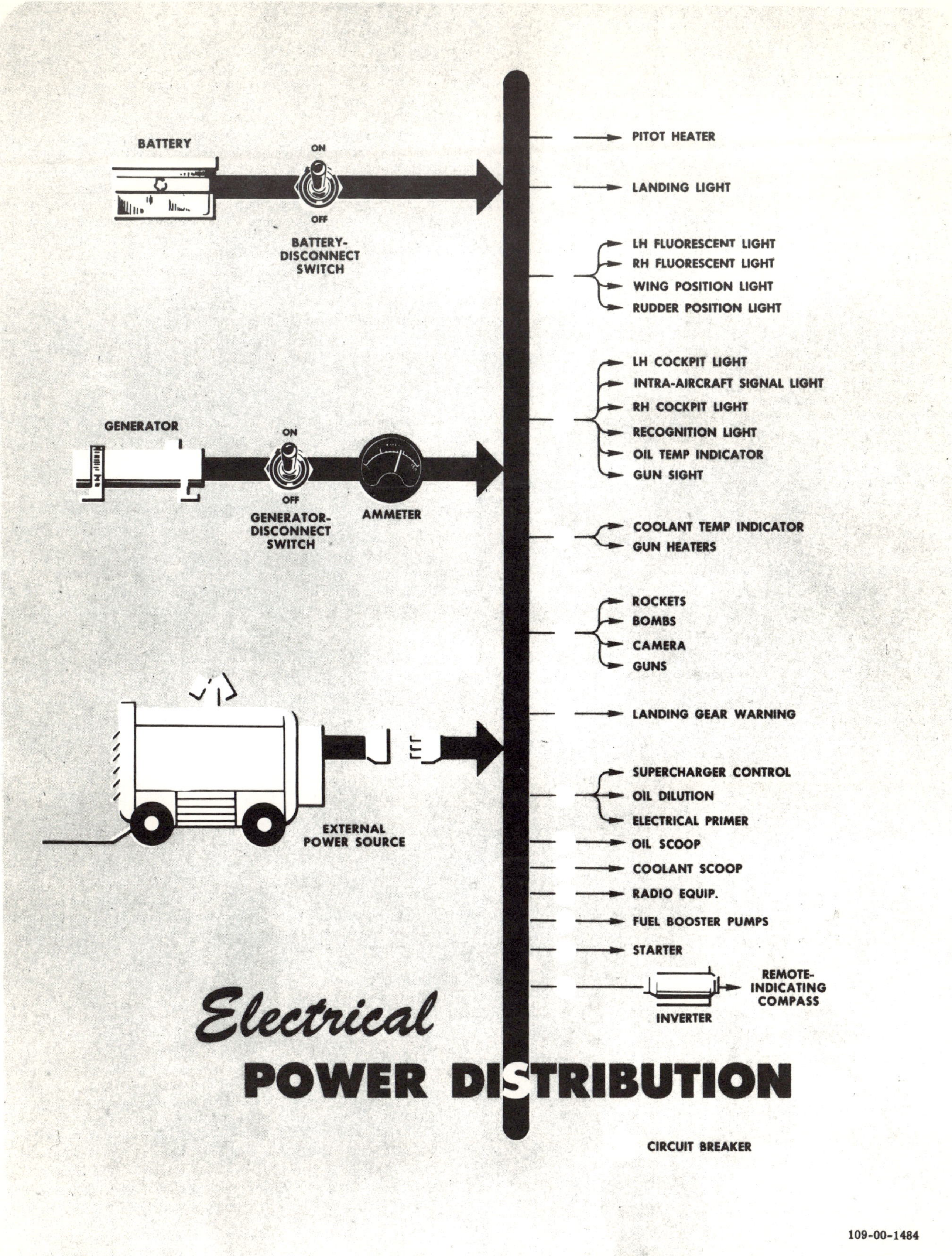

BATTERY
ON
OFF
BATTERY-DISCONNECT SWITCH
GENERATOR
ON
OFF
GENERATOR-DISCONNECT SWITCH
AMMETER
EXTERNAL POWER SOURCE
PITOT HEATER
LANDING LIGHT
LH FLUORESCENT LIGHT
RH FLUORESCENT LIGHT
WING POSITION LIGHT
RUDDER POSITION LIGHT
LH COCKPIT LIGHT
INTRA-AIRCRAFT SIGNAL LIGHT
RH COCKPIT LIGHT
RECOGNITION LIGHT
OIL TEMP INDICATOR
GUN SIGHT
COOLANT TEMP INDICATOR
GUN HEATERS
ROCKETS
BOMBS
CAMERA
GUNS
LANDING GEAR WARNING
SUPERCHARGER CONTROL
OIL DILUTION
ELECTRICAL PRIMER
OIL SCOOP
COOLANT SCOOP
RADIO EQUIP.
FUEL BOOSTER PUMPS
STARTER
REMOTE-INDICATING COMPASS
INVERTER
Electrical
POWER DISTRIBUTION
CIRCUIT BREAKER
109-00-1484

Figure 1-19

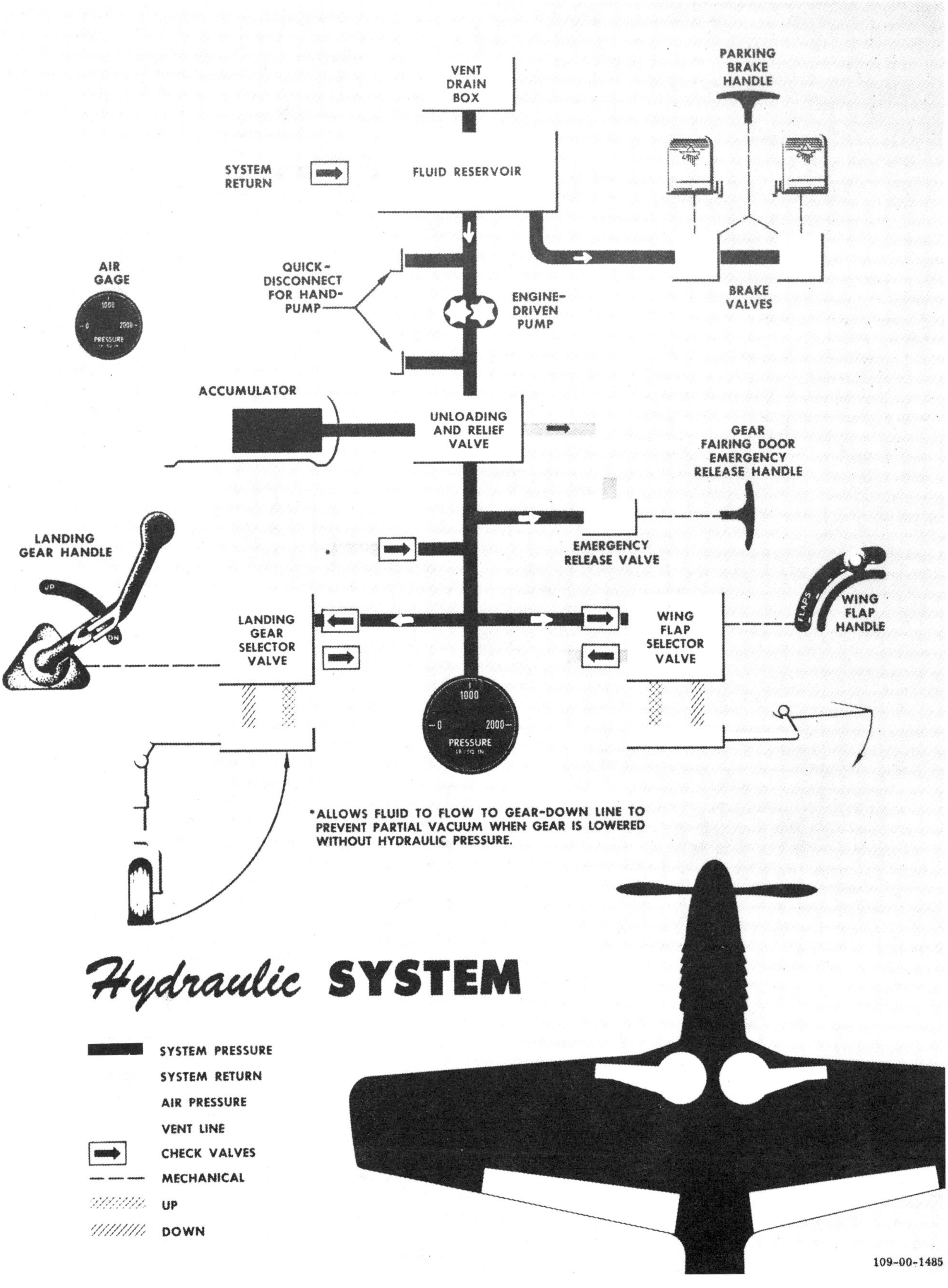
VENT DRAIN BOX
PARKING BRAKE HANDLE
SYSTEM RETURN
FLUID RESERVOIR
AIR GAGE
QUICK-DISCONNECT FOR HAND-PUMP
ENGINE-DRIVEN PUMP
BRAKE VALVES
ACCUMULATOR
UNLOADING AND RELIEF VALVE
GEAR FAIRING DOOR EMERGENCY RELEASE HANDLE
LANDING GEAR HANDLE
EMERGENCY RELEASE VALVE
LANDING GEAR SELECTOR VALVE
WING FLAP SELECTOR VALVE
WING FLAP HANDLE
FLAPS
1000
0
2000
PRESSURE
LB. SQ. IN.
*ALLOWS FLUID TO FLOW TO GEAR-DOWN LINE TO PREVENT PARTIAL VACUUM WHEN GEAR IS LOWERED WITHOUT HYDRAULIC PRESSURE.
Hydraulic SYSTEM
SYSTEM PRESSURE
SYSTEM RETURN
AIR PRESSURE
VENT LINE
CHECK VALVES
MECHANICAL
UP
DOWN
109-00-1485

Figure 1-20

ELECTRICAL POWER SUPPLY SYSTEM CONTROLS AND INDICATOR.

BATTERY-DISCONNECT SWITCH.

A battery-disconnect switch (21, figure 1-5), located on the right switch panel, allows battery power to be supplied to the airplane electrical system. The switch should be in the OFF position when external power is used for starting, to conserve the battery. When the engine is operating and external power is disconnected, the battery-disconnect switch should be placed in the ON position.

GENERATOR-DISCONNECT SWITCH.

The generator-disconnect switch (22, figure 1-5), located on the right switch panel, has ON and OFF positions. With the switch in the OFF position, generator power output is not available to the airplane electrical system. The generator "cut-in" speed is about 1200 rpm and the power output is available when the generator-disconnect switch is at ON. All electrically operated equipment is powered from the generator except the remote-indicating compass, which derives its power from the inverter.

AMMETER.

An ammeter (6, figure 1-5), located on the upper section of the right switch panel, is calibrated for a maximum reading of 150 amperes. The ammeter indicates the amount of current being supplied by the generator. The normal maximum current of 100 amperes should be used only for a short period of time.

HYDRAULIC POWER SUPPLY SYSTEM.

The hydraulic power supply system (figure 1-20) is a closed-center system that incorporates a pressure accumulator and an unloading and relief valve. The system is used to operate the main landing gear and the wing flaps. The hydraulic system operates at a normal pressure of 1050 ($\pm$50) psi. Late airplanes have a 1.2-gallon reservoir which supplies hydraulic fluid to the system; early airplanes have a 1.8-gallon reservoir. The unloading and relief valve is incorporated to prevent excessive pressure. See figure 1-25 for hydraulic fluid grade and specification.

FLIGHT CONTROL SYSTEM.

The ailerons, elevators, and rudder are conventionally operated through push-pull rods and dual cables by a control stick and rudder pedals. Trim tabs on each of the primary surfaces are operable from the cockpit by a cable system. A reverse boost rudder tab is installed on most airplanes. On late airplanes* and on some airplanes modified in service, a 20-pound bobweight has been added to the elevator control system to improve the flight characteristics.

FLIGHT CONTROLS.

CONTROL STICK.

The control stick (figure 4-11) has a conventional-type grip, which incorporates a gun and camera trigger switch. A bomb-rocket release button is located on the top of the stick grip.

CONTROL SURFACE LOCK.

A control surface lock (figure 1-21) at the base of the control stick has two positions. The lock consists of a plunger that snaps into either of two holes. When it is snapped into the lower hole, the ailerons, rudder, and elevators are locked, and the tail wheel is locked in line with the fuselage. When it is snapped into the upper hole, the tail wheel is free to swivel, while the ailerons, rudder, and elevators are locked.

***F-51D-10-NA Airplanes AF44-14653 through -14852 and F-51D-15-NA and subsequent airplanes**

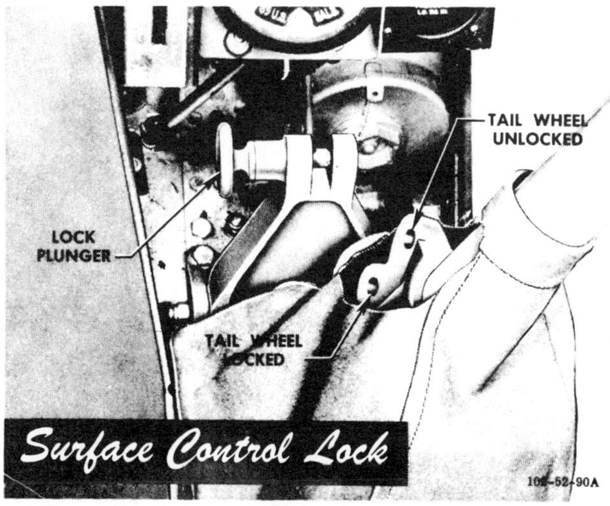

Figure 1-21

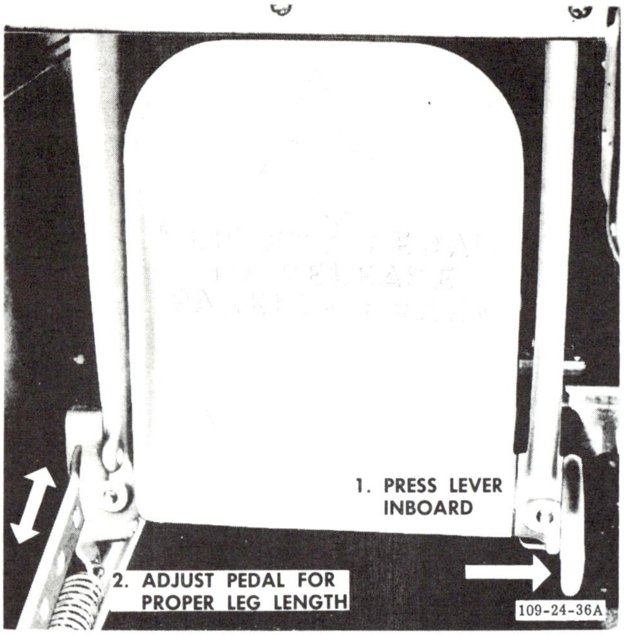

Figure 1-22. Rudder Pedal Adjustment

RUDDER PEDAL ADJUSTMENT LEVER.

The rudder pedals may be adjusted for proper leg length by pushing the pedal adjustment lever (located on the inboard side of each pedal) inboard with the foot and shifting the pedal to the desired position. Releasing the pedal adjustment lever locks the pedal in the new position. (See figure 1-22.) Be sure both pedals are adjusted equally.

TRIM TAB CONTROLS.

ELEVATOR TRIM TAB CONTROL WHEEL. The elevator trim tab control wheel (19, figure 1-4) is located on the left side of the cockpit, below and aft of the throttle quadrant. The control wheel is mounted in a vertical plane and connected to the elevator trim tabs by dual cables. Rolling the wheel forward in the direction of the NH arrow makes the airplane nose-heavy, and rolling the trim wheel in the direction of the TH arrow causes a tail-heavy condition.

RUDDER TRIM TAB CONTROL KNOB. The rudder trim tab control knob (20, figure 1-4) is located horizontally on the left console and is marked "R" and "L" with indicating arrows. A geared pointer indicates the number of degrees the trim tab is moved. Dual cables connect the control knob with the cable drum and actuating screw. A reverse boost action of the trim tab is obtained by a linkage which causes the rudder trim tab to move slightly in the same direction as the rudder, making it necessary to increase rudder pedal pressure to obtain an increase in yaw.

AILERON TRIM TAB CONTROL KNOB. The aileron trim tab control knob (18, figure 1-4) is located horizontally on the left console and is marked "R" and "L" with indicating arrows. A geared pointer indicates the number of degrees the trim tab on the left aileron is moved. (The trim tab on the right aileron is ground-adjustable.) The trim tab control knob is connected to the left aileron trim tab by a chain and dual cables.

WING FLAPS.

The wing flaps are hydraulically actuated, and travel is controlled by a handle in the cockpit. The wing flaps have a total downward travel of 47 degrees. The wing flaps should be full up during taxiing because of the minimum ground clearance afforded. There is no emergency means of lowering the flaps if the hydraulic system fails. However, if malfunction is due to failure of the engine-driven hydraulic pump and all other units are still intact, the hydraulic accumulator has enough pressure to lower the flaps fully, provided the hydraulic pressure gage shows at least 800 psi.

WING FLAP HANDLE.

The wing flap handle (25, figure 1-4) is located on the left side of the cockpit, aft of the console. The handle has six positions: UP, 10°, TAKE OFF 20°, 30°, 40°, and 50°. There is a detent at each of the marked positions, and the flaps hydraulically move to that indicated angle. The maximum angle of flaps permitted is governed by the airspeed. (Refer to Section V.)

LANDING GEAR SYSTEM.

The landing gear system on the airplane is a conventional type, with a steering and locking mechanism provided for the nonretracting tail wheel. The main wheels retract inboard into the belly of the airplane, and fairing doors cover the wheel well and strut openings. In an emergency, a mechanical means of releasing the fairing doors and gear is provided. Yawing the airplane may be necessary to force the gear into a locked position during emergency lowering.

LANDING GEAR SYSTEM CONTROLS.

LANDING GEAR HANDLE.

A landing gear handle (16, figure 1-4), located on the left side of the cockpit just forward of the seat, has two positions: UP and DN. The gear handle positions the landing gear selector valve through a mechanical linkage. The handle is spring-loaded into a detent in its quadrant and must be pulled inboard to be moved from one position to another. The handle is mechanically locked in the down position when the fairing door emergency release handle is pulled (fairing doors open). This prevents accidentally raising the handle while the airplane is on the ground. While the airplane is moving on the ground, the landing gear will retract if the landing gear handle is placed at UP.

LANDING GEAR FAIRING DOOR EMERGENCY RELEASE HANDLE.

A landing gear fairing door emergency release handle (26, figure 1-6) is located to the right of the fuel tank selector handle, forward of the control stick. If the hydraulic system fails, the fairing door emergency release handle may be used to mechanically release the hydraulic pressure and fairing doors after the landing gear handle is placed at DN to release the gear uplocks. Yawing the airplane may be necessary to force the gear into a locked position. The release handle is also used after the airplane is parked, to release the hydraulic pressure and open the fairing doors. Opening the fairing doors mechanically locks the landing gear handle in the down position.

LANDING GEAR SYSTEM INDICATORS.

LANDING GEAR WARNING LIGHTS.

A red light and a green light (35, figure 1-6), located below the center of the instrument panel, are provided for landing gear warning. The lights operate in the following manner:

1. Green light off, red light on, when gear are in any unlocked position, regardless of throttle position, or when gear are up and locked and fairing doors are not fully closed.
2. Green light off, red light off, when gear are up and locked with fairing doors fully closed and throttle forward (beyond minimum cruising power).
3. On early airplanes, green light off, red light on, horn on, when gear are up and locked and throttle is retarded below minimum cruising power.
4. On late airplanes, green light off, red light on, horn on, when gear are in any position other than down and locked and throttle is retarded below minimum cruising power.
5. Green light on, red light off, when gear are down and locked and throttle and doors are in any position.

LANDING GEAR WARNING HORN.

A landing gear warning horn is located between the fuselage skin and armor plate on the left side of the airplane. The horn sounds when the gear are not down and locked and the throttle is retarded below minimum cruising power. A horn cutout button (33, figure 1-6) is located on the front switch panel. The warning horn cuts out when the button is pushed, and the circuit resets itself when the throttle is pushed forward again.

TAIL WHEEL STEERING.

Tail wheel steering features on the airplane are such that with the stick slightly aft of neutral, the tail wheel is locked out of its full-swiveling position and steering through connecting cables is possible up to 6 degrees right or left by use of the rudder pedals. The steering mechanism is released when the stick is pushed forward of the neutral position. The tail wheel cannot be locked after full-swiveling unless the rudder is in the neutral position and the control stick is in the neutral or aft position.

BRAKE SYSTEM.

The main landing gear disk brakes are operated in a conventional manner when the rudder pedals are depressed by toe action. The brake hydraulic system is entirely separate from the general hydraulic system, except that the same reservoir supplies fluid to both systems. A standpipe within this reservoir ensures a reserve for the brake system even though the fluid for the main system may be lost.

PARKING BRAKE HANDLE.

A parking brake handle (38, figure 1-6) is located below the center of the instrument panel. Pulling this handle out, after depressing the brake pedals, locks the brakes. The brake pedals must be released while the parking brake handle is held out. Releasing the brakes is accomplished by depressing both brake pedals.

WARNING

Do not apply brakes after take-off, as the heated disks may fuse or lock together. Sufficient clearance in the wheel wells prevents damage from revolving wheels.

INSTRUMENTS.

The instruments are classified into three groups: flight, engine, and miscellaneous. A yellow line on the panel separates the flight instruments from the other instruments. Suction for the directional gyro, gyro horizon, and turn-and-bank indicator is supplied by an engine-driven vacuum pump. Static pressure for the altimeter, airspeed indicator, and rate-of-climb indicator is taken through a static plate mounted on each side of the fuselage just forward of the stabilizer. Pitot pressure is supplied through a pitot head located beneath the right wing. A remote-indicating magnetic compass is installed. Miscellaneous instruments consist of a hydraulic pressure gage, an ammeter, and a clock. The airplane is not equipped with a stand-by magnetic compass.

EMERGENCY EQUIPMENT.

SIGNAL PISTOL.

A Type M-8 pyrotechnic pistol (2, figure 1-4) is stowed in a canvas holster strapped to the pistol cartridge stowage bag at the left side of the pilot's seat.

Warning

Do not load pistol except when it is in mount, as no safety is provided.

SIGNAL PISTOL MOUNT.

A signal pistol mount is located on the left side of the fuselage, next to the pilot's seat. A cap chained to the pistol mount covers the port when the pistol is not installed.

FIRST-AID KIT.

A first-aid kit (14, figure 1-5) is installed at the right side of the seat. Instructions for use are contained in the kit. The contents should be used only in an emergency when medical aid is not available. Used supplies must be replaced as soon as possible.

CANOPY.

The canopy is a one-piece plastic unit with a metal frame. The canopy is handcrank-operated and slides fore and aft on tracks built onto the fuselage. An external flush button on the right side of the fuselage may be depressed to permit manually sliding the canopy aft for entrance to the airplane. An emergency release handle (operable from inside or outside) is provided.

CANOPY CONTROLS AND INDICATORS.

CANOPY HANDCRANK.

The canopy handcrank (figure 1-23) on early airplanes operates the canopy in the following manner. Pushing on the crank axle engages the clutch, and pulling the crank handle knob inboard disengages the handcrank pin from the locking plate holes. The crank is turned in the desired direction with knob held inboard. The canopy is locked when the locking pin is engaged in the nearest hole. Manually pulling out on the handcrank handle to disengage the clutch permits sliding the canopy to any position desired. On the later airplanes, the handcrank knob has a latch which must be depressed; the canopy can then be cranked to the desired position. Releasing the latch and turning the handcrank until the latch engages, locks the canopy in the desired position.

CANOPY EMERGENCY RELEASE HANDLE.

The canopy emergency release handle (figure 1-24) is located on the right side of the cockpit along the upper longeron, directly forward of the windshield bow. Pulling the handle aft mechanically releases the latches holding the canopy and permits the slip stream to carry the canopy clear of the airplane.

WARNING

Be sure to lower seat and duck head, to avoid being hit by canopy.

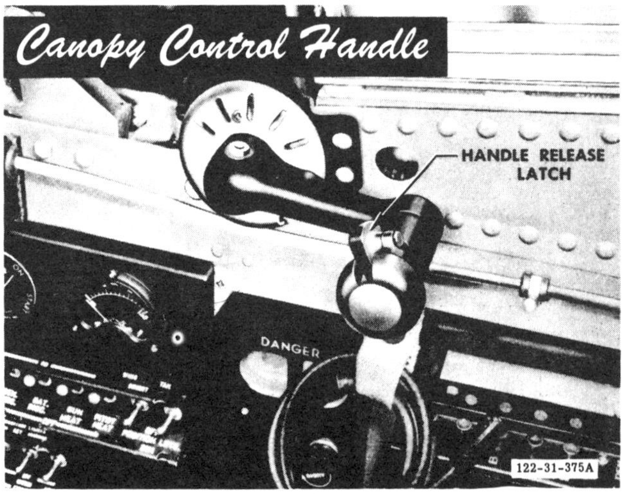

Figure 1-23

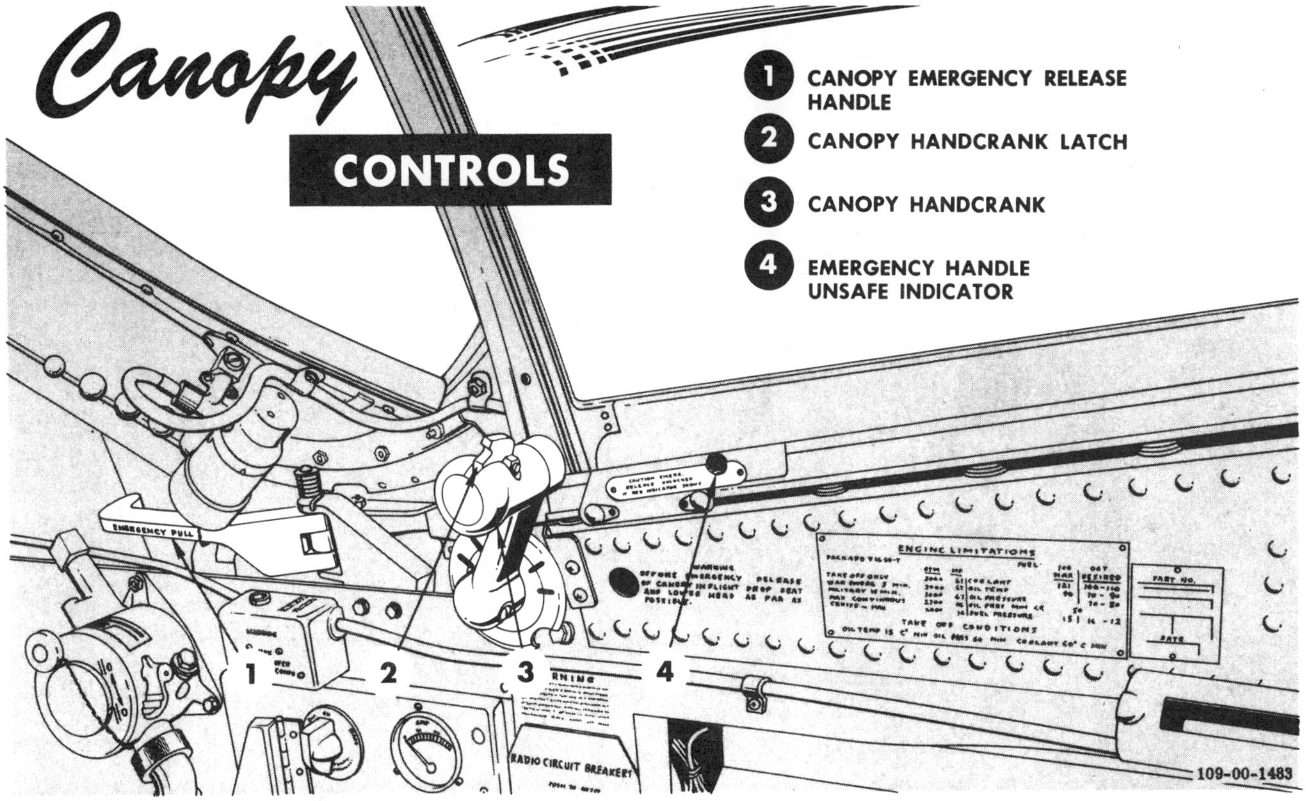

Figure 1-24

Note

> If excessive force is used in securing the canopy before take-off, it may be necessary to crank the canopy back enough to relieve the pressure against the windshield before the emergency release will be effective.

The emergency release handle is also operable from the outside of the airplane to permit ground personnel to open the canopy in an emergency.

CANOPY UNSAFE INDICATORS.

If the canopy emergency release handle or the canopy truck locks are not properly locked and the canopy is unsafe for flight, a red nut will be visible through a cutout on each of the two canopy trucks at the forward end of the canopy frame.

SEAT.

The pilot's seat is adjustable vertically. Two spring-loaded pins, actuated by a lever at the lower right side, snap into any one of nine holes in the seat posts and secure the seat at a desired level. The pilot's parachute is used as a seat cushion, and the kapok-filled seat back cushion may be used as a life preserver. Shoulder straps and a safety belt are attached to the seat and secured by a quick-release safety buckle.

SHOULDER-HARNESS LOCK HANDLE.

On some of the later airplanes, a two-position (locked and unlocked) shoulder-harness lock handle (21, figure 1-4) is located on the left side of the pilot's seat. A latch is provided for positively retaining the handle at either position of the quadrant. When the top of the handle is pressed down, the latch is released and the handle may then be moved freely from one position to another. When the handle is in the UNLOCKED position, the reel harness cable will extend to allow the pilot to lean forward in the cockpit; however, the reel harness cable automatically locks when an impact force of 2 to 3 G is encountered. When the reel is automatically locked in this manner, it remains locked until the handle is moved to LOCKED and then returned to the UNLOCKED position. Rapidly pulling the shoulder harness by hand will not check the automatic locking feature of the inertia reel. When the handle is in the LOCKED position, the reel harness cable is manually locked so that the pilot is prevented from bending forward. If the harness is locked while the pilot is leaning forward, the harness will retract with him as he straightens up, moving in successive locked positions as he moves back against the seat. To unlock the harness, the pilot must be able to lean back enough to relieve the tension on the lock. Therefore, if the harness is locked while the pilot is leaning back hard against the seat, he may not be able

Servicing DIAGRAM

1 . . . . COOLANT HEADER TANK
2 . . . . AFTERCOOLER HEADER TANK
3 . . . . BATTERY (LATE AIRPLANES)
4 . . . . OIL TANK
5 . . . . HYDRAULIC RESERVOIR
6 . . . . LH WING FUEL TANK (RH SIMILAR)
7 . . . . BATTERY (EARLY AIRPLANES)
8 . . . . FUSELAGE FUEL TANK
9 . . . . OXYGEN FILLER

109-00-1480

Figure 1-25

to unlock the harness without first releasing it momentarily at the safety belt (or releasing the harness buckles, if desired). The LOCKED position is used only during aerobatics and flight in rough air, or as an added precaution when a crash landing is anticipated.

CAUTION

All switches not readily accessible with the harness locked should be properly positioned before the harness lock handle is moved forward to the LOCKED position.

AUXILIARY EQUIPMENT.

Information pertaining to the description and operation of the following auxiliary equipment is included in Section IV: heating and ventilating, defrosting, communication, lighting, oxygen, anti-G suit, and armament (guns, bombs, rockets, and sight).

Normal Procedures

SECTION II

STATUS OF THE AIRPLANE.

FLIGHT RESTRICTIONS.

Detailed airplane and engine limitations are listed in Section V.

FLIGHT PLANNING.

From the operating data contained in Appendix I, determine fuel consumption, correct airspeed, and power settings necessary to accomplish the intended mission. The Appendix data will enable you to properly plan your flight so that you can obtain the best possible performance from your airplane.

WEIGHT AND BALANCE.

Refer to Section V for weight and balance restrictions. Refer to Handbook of Weight and Balance Data T.O. No. 1-1B-40 for loading information. Before each mission, make the following checks:

1. Check take-off and anticipated landing gross weight and balance.

2. Check that fuel, oil, armament, and special equipment carried are sufficient for the mission to be accomplished.

3. Check that weight and balance clearance (Form F) is satisfactory.

ENTRANCE.

The cockpit can be entered from either side of the airplane. (See figure 2-1.) However, a spring-loaded flush handhold is provided below the canopy frame on the left side of the fuselage. A marked reinforced section of the wing fillet is used as a step during entry.

Caution

Do not use trailing edge of wing flap or extreme edge of fillet as a step, as damage will result to these units.

Entering Airplane

Enter airplane on left side.

CAUTION:

Avoid stepping on unsupported areas of fillet or wing flaps.

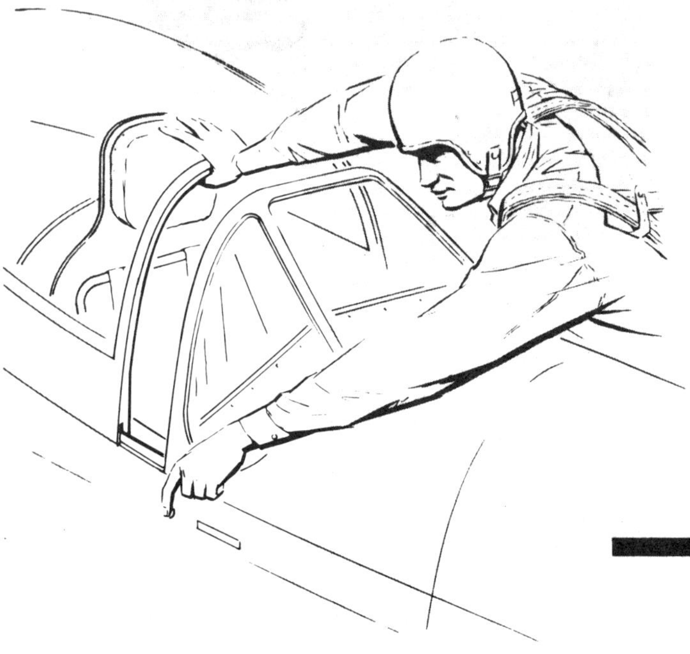

If canopy is closed, reach over windshield and depress release button below windshield frame; then slide canopy open.

NOTE:

After canopy is open sufficiently to reach inside cockpit, use handcrank to run canopy fully open.

109-00-1508

Figure 2-1

EXTERIOR *Inspection* ✓✓✓

NOTE: During this preflight inspection, check entire airplane for wrinkles, loose rivets, dents, and loose access doors.

Starting at the cockpit, make the following checks . . . .

1 COCKPIT:

- Ignition and battery-disconnect switches ***OFF.***
- Check Form 1 for status of airplane and servicing.
- Verify that airplane has been serviced with required quantities of fuel, oil, and hydraulic fluid.
- Trim tabs neutral.
- Compare flap handle position with wing flaps.
- Canopy manually checked for security.

2 LEFT WING SECTION:

- Trailing edge of flaps, wing tips.
- Control surfaces for condition.
- Wing tip and position light.
- Leading edge and gun bay doors.
- Drop tank level, cap secure.
- Fuel tank level, cap secure.
- Wheel well for leaks, gear assembly, downlocks, strut inflation approximately 3-7/16 inches.
- Tire for inflation, wear, and slippage.
- Landing light.
- Wheels chocked.

3 ENGINE SECTION:

- Cowl for security.
- Exhaust stock plugs removed.
- Air scoops for obstructions.
- Propeller for nicks and excessive oil.

4 RIGHT WING SECTION:

- Tire for proper inflation, wear, and slippage.
- Wheel well for leaks, check gear assembly and downlocks, strut inflation at approximately 3-7/16 inches, accumulator air pressure 400 (±25) psi.
- Fuel tank level, cap secure.
- Drop tank level, cap secure.
- Wing leading edge for condition, gun bay doors secure.
- Pitot tube cover removed.
- Wing tip and position light.
- Control surfaces for condition.
- Trailing edge and flaps for condition.

5 RIGHT FUSELAGE:

- Coolant door for cracks, leaks, and position.
- Oil shutter for position and condition.
- Static-pressure vents unobstructed.
- Radio antennas.

6 TAIL SECTION:

- Control surfaces for condition.
- Trim tab position.
- Tail wheel for strut inflation and tire slippage.
- Position light.

7 LEFT FUSELAGE:

- Fuel cap secure.
- Static-pressure vents unobstructed.

109-00-1507

Figure 2-2

BEFORE EXTERIOR INSPECTION.

Check Form 1 for engineering status and make sure airplane has been properly serviced. If at a strange field, check cooling system before flight. Before removing coolant cap, let cool for at least one hour. See figure 1-25 for complete servicing data. Prior to the exterior inspection, make the following safety checks:

1. Landing gear handle DN.
2. Battery-disconnect switch OFF.
3. Ignition switch OFF.
4. Gun safety switch OFF.
5. Bomb arming switch OFF.
6. Rocket release switch OFF.
7. Bomb-rocket selector switch SAFE.

EXTERIOR INSPECTION.

Exterior inspection should be accomplished as shown in figure 2-2.

ON ENTERING COCKPIT.

INTERIOR CHECK (ALL FLIGHTS).

Note

This procedure is arranged in a clockwise direction around the cockpit to minimize the necessary motions and still check each item thoroughly.

1. Fasten safety belt and shoulder harness. Check operation of shoulder-harness lock.
2. Adjust seat level to obtain full travel of rudder pedals in extreme position.
3. Adjust rudder pedals for proper leg length to obtain full brake control. Press foot against lever on inner side of each pedal.
4. Unlock control lock at base and just forward of control stick by pulling plunger on side of lock.
5. Check controls for free and proper movement, watching control surfaces for correct response.
6. Wing flap handle full UP.
7. Carburetor ram-air control lever at UNRAMMED FILTERED AIR for all ground operation.
8. Carburetor hot-air control lever NORMAL (late airplanes).
9. Landing light switch OFF.
10. Windshield defroster and hot-air control knobs OFF.
11. Check fuel quantity gages.
12. Throttle one inch open (in START position on late airplanes).
13. Mixture control at IDLE CUTOFF.
14. Propeller control full INCREASE.
15. Friction locks on throttle quadrant adjusted for friction.
16. Gun sight gyro motor switch ON (K-14A only).
17. Gun sight gyro selector switch at FIXED.
18. Parking brakes set.
19. Supercharger control switch AUTO.
20. Clock set.
21. Gyro instruments uncaged.
22. Altimeter set to field elevation.
23. Note manifold pressure reading (field barometric pressure) for subsequent use during preflight engine check.
24. Fuel shutoff lever ON.
25. Fuel tank selector handle to FUS. TANK. If fuselage tank is not serviced, selector handle to MAIN TANK L. H.
26. Ignition switch OFF.
27. Landing gear fairing door emergency release handle in.
28. Oxygen gage pressure 400 psi. Test oxygen equipment for operation.
29. Check canopy emergency release handle and coolant flap emergency release handle for safetying.
30. Radio and communication equipment switches OFF.
31. Check landing gear warning lights. (Battery-disconnect switch ON temporarily.)
32. Check generator-disconnect switch ON.
33. Gun heater switch OFF.
34. All circuit breakers pushed in.

INTERIOR CHECK (NIGHT FLIGHTS).

In addition to the preceding check, perform the following checks for night operation:

1. Turn on and check all cockpit lights.
2. Turn on position lights.
3. Make sure personal gear includes a flashlight.

STARTING ENGINE.

The following procedure should be used to start the engine. Starting should be accomplished with the airplane on a paved surface and headed into the wind whenever possible. Have fire guard stand at right wing tip for safety.

1. Ignition switch and battery-disconnect switch OFF.
2. Mixture control IDLE CUTOFF.
3. Have ground crew pull propeller through several revolutions.
4. External power supply connected. (Battery-disconnect switch ON if external power supply is not available.)

Note

Use of battery power is considered an emergency measure.

5. Check throttle open approximately one inch (1500 rpm). (To START position on late airplanes.)

CAUTION

To prevent a runaway engine, the throttle should not be advanced beyond one inch for starting. If the throttle is beyond this point, the butterfly position in the carburetor must be visually checked before the engine is started.

6. Oil and coolant radiator air control switches at OPEN until flaps are fully opened; then release switches to OFF.

CAUTION

For all ground operation, oil and coolant flaps should be fully opened to prevent overheating.

7. Check that propeller is clear.
8. Hold starter switch at ON.
9. Ignition switch to BOTH after six blades have passed.

CAUTION

Keep hand on ignition switch ready for emergency shutoff in case of runaway engine. (In case of a runaway engine, the airplane must be tied down for the next run-up.)

10. Fuel booster pump switch to ON (NORMAL on early airplanes).
11. Prime engine on early airplanes three or four strokes when cold, one stroke when hot. On late airplanes, primer switch ON 3 or 4 seconds when cold, one second when hot.
12. When engine starts, move mixture control to NORMAL and release primer switch as engine smooths out. Do not jockey throttle. If engine does not start after turning several revolutions, continue priming.

WARNING

The mixture control should always be in IDLE CUTOFF when the engine is not firing, to prevent excess fuel entering the induction system and causing a fire hazard.

13. Check oil pressure. If it is not at 50 psi within 30 seconds after engine starts, stop engine and investigate.
14. Move battery-disconnect switch to ON after disconnecting external power source.

ENGINE GROUND OPERATION.

After engine starts, place supercharger control switch at HIGH; then warm up engine at 1300 rpm until oil temperature shows a definite increase and oil pressure remains steady when additional throttle is applied. The following checks should then be made:

WARNING

Do not exceed 2200 rpm in high blower on the ground, as this will tend to cause detonation.

1. Fuel system check—rotate fuel tank selector handle and check fuel pressure for proper operating range of each tank. Fuel booster pump switch must be ON (EMERG. on early airplanes). If drop tanks are installed, check fuel flow from each. Position fuel tank selector handle at MAIN TANK L. H. for take-off.
2. Radiator air outlet flaps—move coolant flap and oil radiator air control switches to OPEN and CLOSE positions and have outside observer verify their operation. Hold switches at OPEN until radiator air outlet flaps are fully open; then release switches to OFF.
3. Check oil, coolant, and fuel gages for proper indications. Place supercharger control switch at AUTO.
4. Ignition system check—at 700 rpm, turn ignition switch OFF momentarily. If engine does not cease firing completely, shut down engine and warn all personnel to keep clear of propeller.

CAUTION

Perform this check as rapidly as possible, to prevent severe backfire when ignition switch is returned to BOTH.

5. Propeller check—with propeller control in full INCREASE, set throttle to obtain 2300 rpm. Move propeller back to DECREASE position to note maximum drop of 300 rpm. Return control to INCREASE.

6. Simmond's regulator check—watch manifold pressure during propeller check. Manifold pressure should remain constant within one in. Hg.

7. Supercharger check—at 2300 rpm, place supercharger control switch at HIGH; there should be at least a 50 rpm drop. Return supercharger control switch to AUTO.

8. Deleading spark plugs—should prolonged ground operation be necessary, such as for checking engine condition or performing numerous preflight checks, run engine at 61 in. Hg manifold pressure and 3000 rpm for one continuous minute prior to take-off.

Caution

Do not exceed 40 in. Hg during ground run-up without having tail tied down, because of the possibility of airplane nosing over.

GROUND TESTS.

Check operation of wing flaps. Turn on necessary communication equipment and ascertain that signals are audible and clear. Check instruments in proper ranges.

TAXIING.

Use the following procedure during taxiing:

1. Remove chocks and release parking brake. Let airplane roll forward slightly, and check brakes.

CAUTION

Never allow taxi speed to build up before checking brakes.

2. Steer a zigzag course to obtain an unobstructed view.

3. Taxi with stick slightly aft of neutral to prevent excessive loads on tail wheel and to lock tail wheel. In the locked position, the tail wheel may be turned 6 degrees right or left with the rudder pedals. For sharp turns, push stick slightly forward of neutral position to allow full-swiveling action of tail wheel.

4. Use brakes as little as possible, to prevent overheating.

CAUTION

To avoid excessive use of brakes, taxi at idle rpm.

5. Upon reaching take-off position, stop airplane at right angles to runway so that approaching airplanes may be seen.

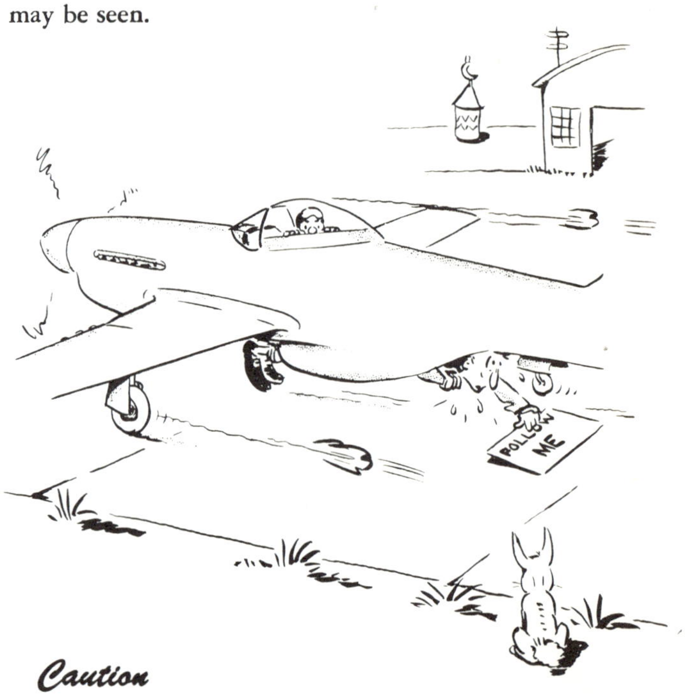

Caution

Taxi cautiously to avoid damage from objects which the tires or propeller might pick up and throw against the radiator air outlet flaps.

BEFORE TAKE-OFF.

PREFLIGHT AIRPLANE CHECK.

Before take-off, check safety belt fastened and shoulder harness unlocked; then check:

1. Primary Controls:
   Check surface controls for free movement.
2. Instruments and Switches:
   Altimeter set.
   Directional gyro set.
   Gyro horizon set.
   All instrument readings in desired ranges.
   All switches and controls at desired positions.
3. Fuel System:
   Check fuel tank selector handle on MAIN TANK L. H. Be sure selector is in detent. (Refer to Section VII for instructions concerning fuel sequence during flight.)
   Fuel booster pump switch at ON (EMERG. on early airplanes).
   Primer switch OFF (locked on early airplanes).
4. Flaps:
   Flaps set for take-off (UP for normal take-off).
5. Trim:
   Trim tabs set for take-off:

	FUSELAGE TANK 0 TO 25 GAL	FUSELAGE TANK FULL (65 GAL)
Rudder	6 deg R	6 deg R
Elevator	0 deg	2 to 4 deg nose-heavy
Aileron	0 deg	0 deg

6. Check all circuit breakers in.
7. Check that cockpit enclosure is locked and that canopy emergency release handle is safetied.

PREFLIGHT ENGINE CHECK.

Note

Tap instrument panel while performing checks requiring rpm readings, to prevent tachometer sticking.

1. Check propeller at full INCREASE.
2. Power check—advance throttle to obtain 2300 rpm. At this rpm, the manifold pressure should read ½ in. Hg less than field barometric pressure within ±½ in. Hg.

Note

Manifold pressure in excess of field barometric pressure indicates that the engine is not producing maximum power and should be checked.

3. Ignition system check—at 2300 rpm, with propeller in full INCREASE, move ignition switch from BOTH to L, back to BOTH, then to R, and back to BOTH. Let engine speed stabilize at BOTH between checks. A maximum drop of 100 rpm is allowable for the right magneto and 120 rpm drop for the left magneto. If rpm drop is more than allowable, spark plugs will have to be deleaded. (Refer to Engine Ground Operation in this section.)
4. Idle speed check—idle engine at 650 to 700 rpm with throttle against idle stop.
5. Acceleration and deceleration check—with mixture control at NORMAL, advance throttle from idle to 2300 rpm. Engine should accelerate and decelerate smoothly with no tendency to backfire.
6. Set throttle for 1500 rpm for best cooling during prolonged ground operation.
7. Carburetor ram-air control lever at RAM AIR. (UNRAMMED FILTERED AIR or carburetor hot-air control lever at HOT AIR only if required.)

CAUTION

Anticipate longer take-off run if HOT AIR position is used.

8. Check mixture control at NORMAL.
9. Check supercharger control switch at AUTO.

10. Oil and coolant radiator air control switches at AUTOMATIC.

Caution

If coolant temperature exceeds 100°C, place coolant radiator air control switch in open position until air-borne.

CAUTION

If coolant temperature exceeds limits and/or the coolant relief valve pops off, the engine must be immediately shut down and inspected for coolant leaks.

11. If it is necessary to wait long before take-off, recheck magnetos to see if any spark plug leading is present.

TAKE-OFF.

Plan your take-off according to the following variables affecting take-off technique: gross weight, wind, outside air temperature, type of runway, and height and distance of the nearest obstacles. See figure A-4 for required take-off distances.

NORMAL TAKE-OFF.

In order to perform a take-off within the distance specified in the Take-off Distances charts (figure A-4), the following procedure must be used:

1. Be sure take-off area is clear and check final approach for aircraft.
2. Release brakes and line up for take-off.
3. Advance throttle smoothly and steadily to Take-off Power.

Note

It is recommended that 61 in. Hg and 3000 rpm be used for all take-offs and that this power setting be reached as quickly as possible after the take-off run is started. Do not jam throttle forward, as torque will cause loss of control of airplane.

4. If rough engine occurs during take-off run, immediately throttle back 4 or 5 in. Hg manifold pressure to complete take-off if conditions permit. Throttling back tends to decrease the intensity of detonation or preignition and minimizes the chances of engine failure. If this condition occurs on take-off, the spark plugs must be changed before the next flight.

Figure 2-3 (Sheet 1 of 2)

5. Do not attempt to lift tail too soon, as this increases torque action. Pushing the stick forward unlocks the tail wheel, thereby making steering difficult. The best take-off procedure is to hold the tail down until sufficient speed for rudder control is attained and then to raise the tail slowly.

MINIMUM-RUN TAKE-OFF.

To accomplish a minimum-run take-off (figure 2-3), lower flaps 15 to 20 degrees. Keep airplane in a three-point attitude and allow it to fly itself off ground in this position. As soon as air-borne, allow airspeed to build up and climb out at 100 mph. Retract landing gear when airspeed reaches a safe value. Raise flaps above 200 feet altitude.

CROSS-WIND TAKE-OFF.

The following procedure is recommended for a cross-wind take-off:

1. Advance throttle to Take-off Power.
2. Hold tail down until sufficient speed is attained to ensure positive rudder control. Speed should be slightly greater than for normal take-off.
3. Apply sufficient aileron control to keep wings level or even to effect a slightly wing-low attitude into wind.
4. Keep airplane firmly on runway until speed is sufficient to make a smooth, clean break.
5. After becoming air-borne, crab into wind enough to counteract drift.

NIGHT TAKE-OFF.

Night take-off procedure is the same as that for daylight operation. However, a thorough knowledge of switch and light location is essential. The following additional checks are recommended for night take-off:

1. Turn cockpit lights low.
2. Tune radio carefully and loud, as it will fade during take-off and flight.
3. Hold airplane steady on a reference point during take-off run.

AFTER TAKE-OFF.

1. As soon as airplane is definitely air-borne, retract landing gear by pulling landing gear handle inboard and up. Check position of gear by warning lights.
2. On minimum run take-off, when sufficient airspeed is attained and all obstacles are cleared, raise flaps to full up position. No sink is noticeable when flaps are raised.
3. Check coolant and oil temperatures and oil pressure.

WARNING

Do not apply brakes after take-off to stop rotation of wheels, as brake disks may seize.

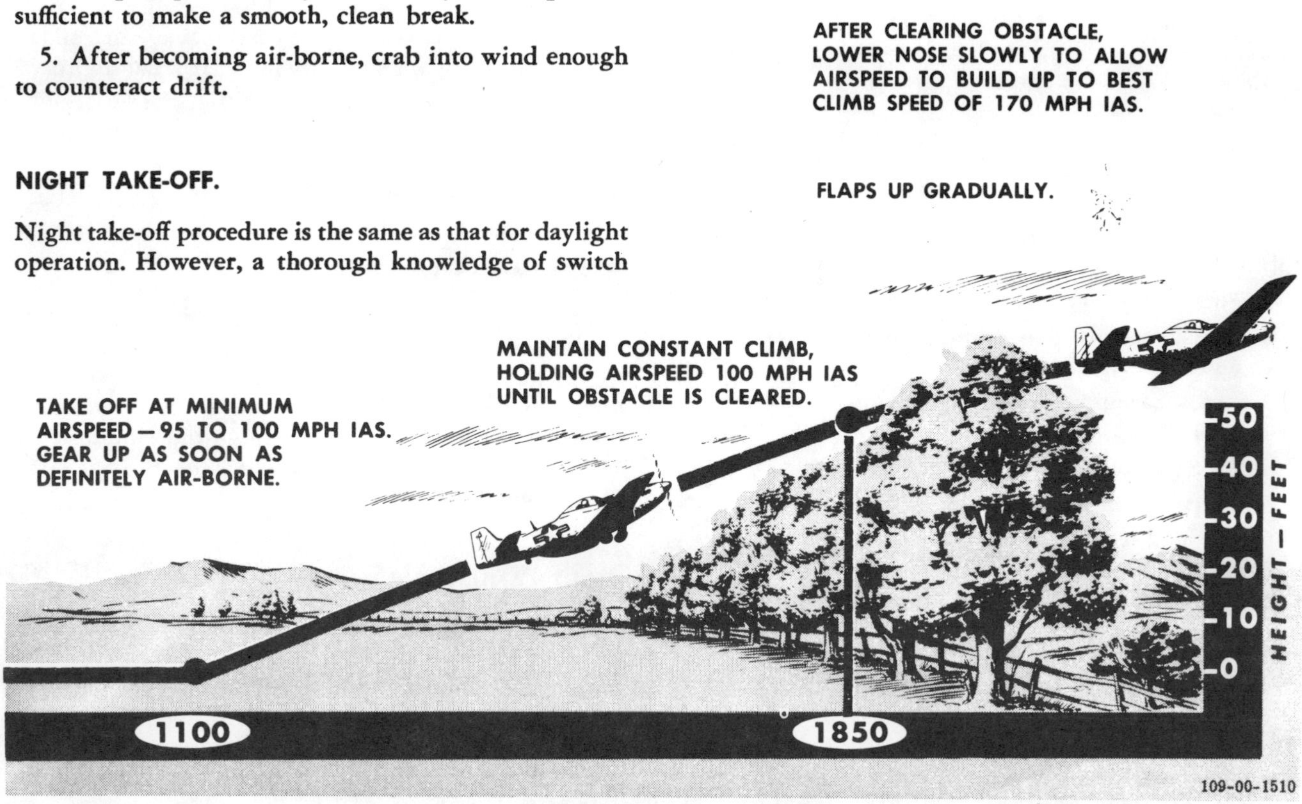

Figure 2-3 (Sheet 2 of 2)

CLIMB.

1. Allow airspeed to build up to 170 mph for normal climb.

2. Check coolant and oil temperatures and oil pressure during flight.

3. Refer to climb charts (figures A-5 and A-6) for power settings, recommended airspeed, rate of climb, and fuel consumption.

FLIGHT CHARACTERISTICS.

Refer to Section VI for all data on flight characteristics.

SYSTEMS OPERATION.

Refer to Section VII for information on systems operation.

DESCENT.

Before descent, turn windshield defroster control knob ON. Descent may be carried out at any safe speed down to the recommended margin of about 25 percent above stalling speed. With the landing gear and flaps up, the glide is fairly flat with the nose very high. Forward visibility is poor in this condition, and in traffic areas, a series of mild "S" turns should be employed to prevent possible collision. Lowering either the flaps or landing gear, or both, greatly increases the gliding angle and the rate of descent.

PRE-TRAFFIC-PATTERN CHECK.

Before entering the traffic pattern, accomplish the following:

1. Fuel tank selector handle on fullest internal tank.

2. Check that fuel booster pump switch is ON (NORMAL on early airplanes).

3. Check carburetor ram- and hot-air control levers as needed.

4. Mixture control at NORMAL.

5. Propeller set at 2700 rpm.

6. Oil and coolant radiator air control switches AUTOMATIC.

7. Clean out engine at 3000 rpm and 61 in. Hg for one minute.

TRAFFIC-PATTERN CHECK.

Traffic-pattern procedure and check are shown in figure 2-4.

LANDING.

NORMAL LANDING.

In order to obtain the results stated in the Landing Distances chart (figure A-7), accomplish the approach and landing procedures outlined in figure 2-4. For a normal landing, plan your approach so that you are over the edge of the field at 120 mph. Use a continuous back pressure on the stick to obtain a tail-low attitude for actual touchdown. Because of the wide landing gear and locked tail wheel, landing roll characteristics are excellent on this airplane. Minimize use of brakes during ground roll. At completion of landing roll, clear runway as soon as possible. Refer to Section III for information regarding emergency landings.

CROSS-WIND LANDING.

In accomplishing a cross-wind landing, maintain an airspeed slightly higher than for a normal approach. Either use the slip method by lowering the upwind wing or crab into the wind to align flight path with runway. Align airplane with runway at touchdown and maintain direction control with rudder. Minimize use of brakes during landing roll. As soon as practical, clear the runway and stop.

HEAVY-WEIGHT LANDING.

If a heavy-weight landing is to be attempted, maintain an airspeed approximately 20 mph over normal approach speed. Power should be used if a flat approach is made. Flare out smoothly and reduce power until touchdown is effected, and then cut off power completely. Do not use a full stall landing. Complete landing roll as in normal landing.

MINIMUM-RUN LANDING.

Minimum-run landings may be accomplished in either of two ways. If no obstacle is present, lower flaps fully and make a flat power-on approach. Hold airspeed to lowest possible safe limit. When in position, close throttle completely. For a minimum-run landing over an obstacle, lower flaps fully and close throttle completely when sure of clearing obstacle.

NIGHT LANDING.

The same techniques and procedures used for day landings should be used. If landing in thick haze or fog, avoid use of landing light, as reflection from the light impedes vision and may distort depth perception. Use the landing light as little as necessary while on the ground. After stopping, clear runway as soon as possible.

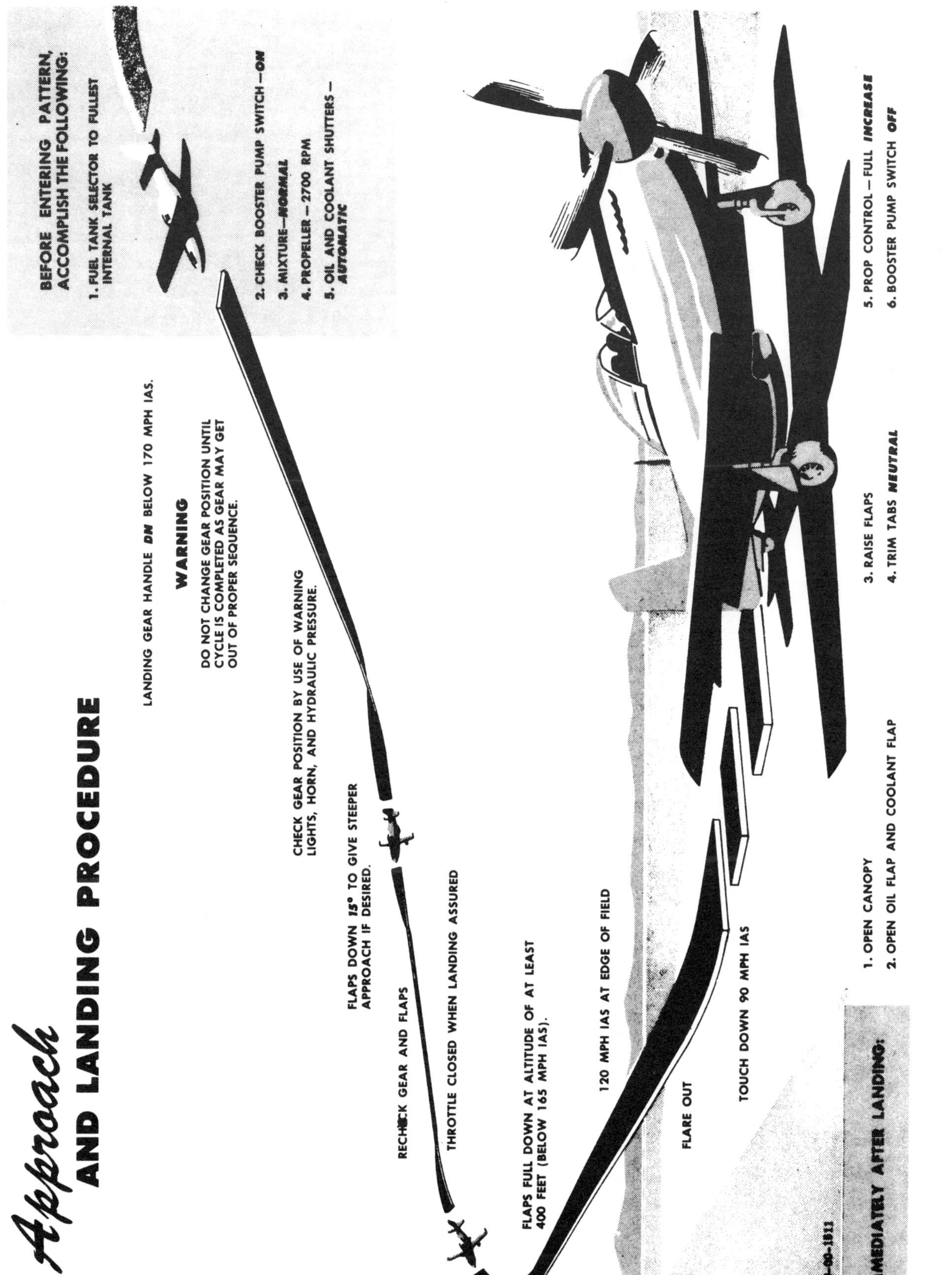
Approach
AND LANDING PROCEDURE
BEFORE ENTERING PATTERN, ACCOMPLISH THE FOLLOWING:
1. FUEL TANK SELECTOR TO FULLEST INTERNAL TANK
2. CHECK BOOSTER PUMP SWITCH—ON
3. MIXTURE—NORMAL
4. PROPELLER—2700 RPM
5. OIL AND COOLANT SHUTTERS—AUTOMATIC
LANDING GEAR HANDLE DN BELOW 170 MPH IAS.
WARNING
DO NOT CHANGE GEAR POSITION UNTIL CYCLE IS COMPLETED AS GEAR MAY GET OUT OF PROPER SEQUENCE.
CHECK GEAR POSITION BY USE OF WARNING LIGHTS, HORN, AND HYDRAULIC PRESSURE.
FLAPS DOWN 15° TO GIVE STEEPER APPROACH IF DESIRED.
RECHECK GEAR AND FLAPS
THROTTLE CLOSED WHEN LANDING ASSURED
FLAPS FULL DOWN AT ALTITUDE OF AT LEAST 400 FEET (BELOW 165 MPH IAS).
120 MPH IAS AT EDGE OF FIELD
FLARE OUT
TOUCH DOWN 90 MPH IAS
IMMEDIATELY AFTER LANDING:
1. OPEN CANOPY
2. OPEN OIL FLAP AND COOLANT FLAP
3. RAISE FLAPS
4. TRIM TABS NEUTRAL
5. PROP CONTROL—FULL INCREASE
6. BOOSTER PUMP SWITCH OFF

Figure 2-4

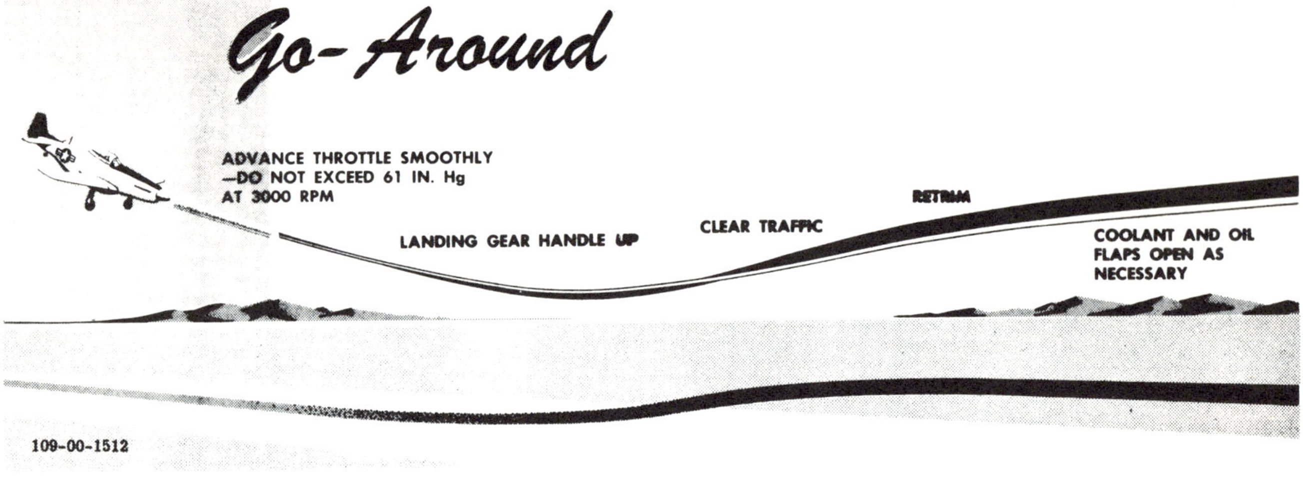

Figure 2-5 (Sheet 1 of 2)

GO-AROUND.

If a go-around is necessary (figure 2-5), use the following procedure:

1. Open throttle smoothly; do not exceed 61 in. Hg at 3000 rpm.
2. Maintain wings level and nose straight.
3. Landing gear handle UP.
4. Raise flaps slowly when at least 200 feet above ground.

AFTER LANDING.

After landing, clear the runway as soon as possible and perform the following:

1. Set throttle at 1000 rpm.
2. Open canopy.
3. Oil and coolant radiator air control switches at OPEN. Release switches to OFF when flaps are fully open.
4. Raise wing flaps completely.
5. Set trim tabs at neutral.
6. Set propeller control at full INCREASE.
7. Fuel booster pump switch OFF.

POSTFLIGHT ENGINE CHECK.

After the last flight of the day, make the following checks:

Note

While performing checks requiring rpm reading, it may be necessary to tap the instrument panel to prevent tachometer sticking, especially in cold weather.

1. Check propeller control at full INCREASE.
2. Ignition switch check—at 700 rpm, turn ignition switch OFF momentarily. If engine does not cease firing completely, shut down engine and warn personnel to keep clear of propeller until discrepancy is corrected.

CAUTION

Perform this check as rapidly as possible, to prevent severe backfire when ignition switch is returned to BOTH.

3. Idle speed and mixture check—with throttle against idle stop, the engine should idle at 650 to 700 rpm. When engine idle speed is stabilized, slowly move mixture control toward IDLE CUTOFF and note any change in rpm. The rpm should flick up very slightly, then decrease. A large noticeable rise in rpm indicates that the mixture is too rich. Absence of the slight flick up but a decrease of rpm indicates too lean a mixture. Excessively rich or lean mixtures increase cylinder head temperature and promote spark plug fouling. Return mixture control to NORMAL before engine cuts out.
4. Power check—advance throttle until rpm is 2300. At this rpm, the manifold pressure should read ½ in. Hg less than field barometric pressure within ±½ in. Hg.

Note

Manifold pressure in excess of field barometric pressure indicates that engine is not producing maximum power and should be checked.

STOPPING ENGINE.

When a cold-weather start is anticipated, dilute oil as required by the lowest expected temperature. For oil dilution instructions, refer to Section IX.

Figure 2-5 (Sheet 2 of 2)

1. Parking brakes set.
2. Advance throttle to 1500 rpm and run until temperatures stabilize to prevent hot spots.
3. Mixture control to IDLE CUTOFF.

WARNING

Do not advance throttle after moving mixture control to IDLE CUTOFF, to prevent runaway engine at next start.

4. Ignition switch OFF after engine stops firing.
5. Fuel shutoff lever OFF.
6. Fairing door emergency release handle pulled out and down. (When fairing doors are open, landing gear handle is mechanically locked in the DN position.)
7. Radio off.
8. All electrical switches OFF.
9. Battery-disconnect switch OFF. Leave generator-disconnect switch ON.

BEFORE LEAVING AIRPLANE.

1. Have wheels chocked; then release brakes.
2. Controls locked.

CAUTION

Use upper hole in control stick when airplane is to be towed, so that tail-wheel will be free to swivel.

3. Carburetor ram-air control lever at UNRAMMED FILTERED AIR.
4. Complete Form 1.
5. Close canopy.

Emergency Procedures

SECTION III

ENGINE FAILURE.

Engine failures fall into two main catagories: those occuring instantly, and those giving ample warning. The instant failure is rare and usually occurs only if ignition or fuel flow completely fails. Most engine failures are gradual and afford the alert pilot ample indication that he may expect a failure. An extremely rough-running engine, loss of oil pressure, excessive coolant temperature under normal flight conditions, loss of manifold pressure, and fluctuating rpm are indications that a failure may occur. When indications point to an engine failure, the pilot should land immediately.

ENGINE AIR RESTART.

If the engine fails in flight and you have sufficient altitude, you may attempt a restart, provided the engine did not fail for obvious mechanical reasons. Unless the engine seizes or internal structural failure occurs, the propeller will windmill even at minimum glide speed. Should airspeed inadvertently drop to a value where the propeller ceases to rotate, the airplane should be nosed down to regain additional speed. In nearly all cases, the propeller will start to rotate again. If necessary, the starter may be used to turn the engine over. All unnecessary electrical equipment should be turned off before the starter is used. Use normal starting procedure after checking fuel tank selector handle on fullest tank.

ENGINE FAILURE DURING TAKE-OFF RUN.

The chances of engine failure during take-off can be greatly reduced if the engine is run up carefully and checked thoroughly beforehand. If engine failure occurs during take-off run before the airplane leaves the ground, proceed as follows:

1. Close throttle completely.
2. Apply brakes as necessary to effect a quick stop.
3. If doubt exists as to whether airplane can be brought to a safe stop on runway, ignition switch OFF and fuel shutoff lever OFF.
4. If insufficient runway remains for a safe stop or obstacles cannot be avoided, jettison external load and move landing gear handle UP.
5. Roll canopy back or pull canopy emergency release handle.
6. Shoulder harness locked.
7. After stopping, get out of airplane as soon as possible, and remain outside.

ENGINE FAILURE DURING TAKE-OFF (AIRPLANE AIR-BORNE).

Move mixture control to RICH if engine begins to fail. If engine fails completely immediately after take-off (figure 3-1), act quickly as follows:

1. Lower nose at once, so that airspeed does not drop below stalling speed.

2. Pull bomb-tank salvo levers to release tanks or bombs.

3. Release sliding canopy by pulling the canopy emergency release handle.

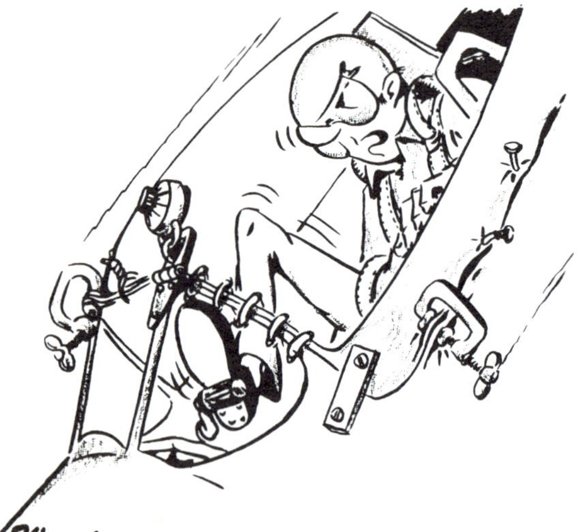

Warning

If excessive force was used to secure canopy before take-off, it may be necessary to crank canopy back to relieve pressure against windshield before emergency release is effective.

WARNING

Before emergency release of canopy in flight, drop seat and lower head as far as possible to avoid being struck by canopy.

4. If there is a reasonable doubt as to condition of terrain on which you are forced to land, or if there is a probability of airplane nosing over or overrunning available landing area, move landing gear handle UP.

5. If time permits, place wing flap handle full DN.

6. Move mixture control to IDLE CUTOFF and turn ignition switch OFF.

7. Move fuel shutoff lever to OFF.

8. Turn battery-disconnect switch OFF.

9. Shoulder harness locked. (Cut all switches before locking harness.)

10. Land straight ahead, changing direction only enough to miss obstructions.

11. After landing, get out of airplane as quickly as possible and remain outside.

109-00-1519

Figure 3-1

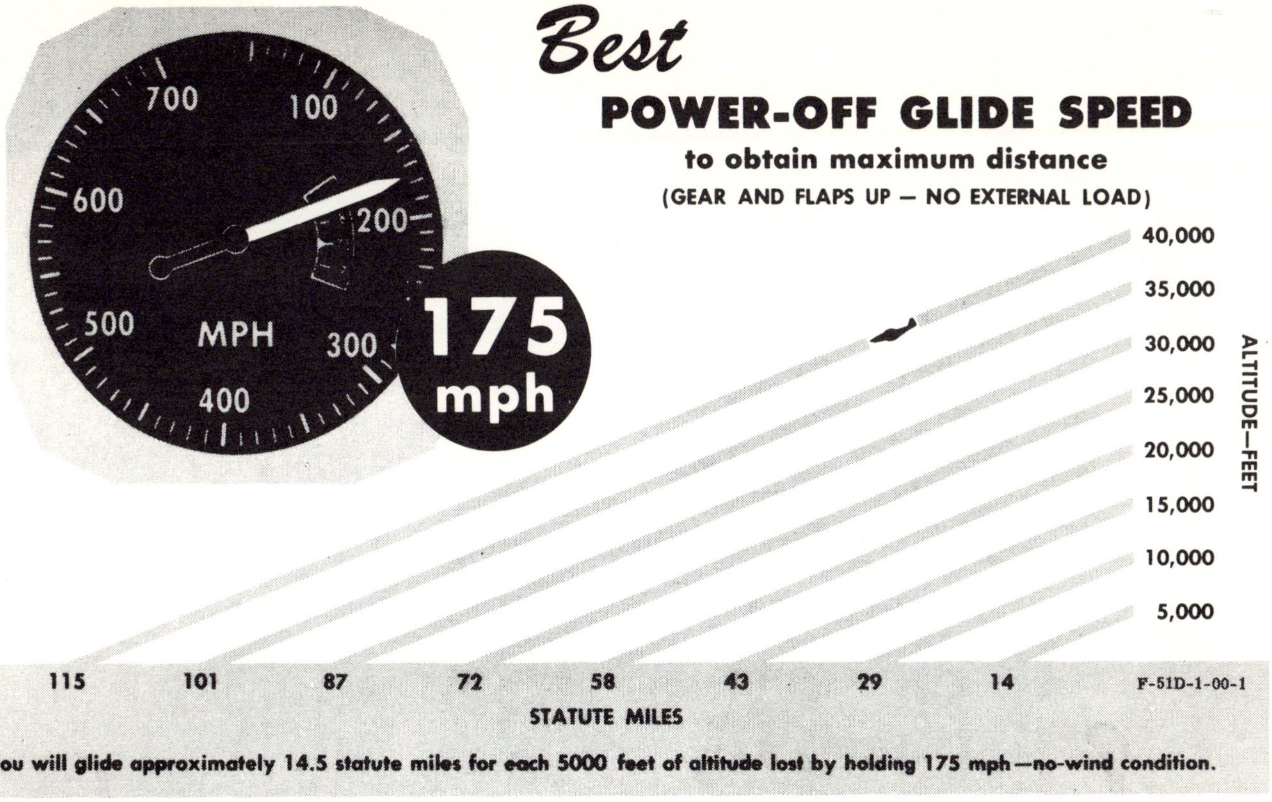

Figure 3-2

ENGINE FAILURE DURING FLIGHT.

If the engine begins to fail during flight, immediately move the mixture control to RICH. If the engine fails during flight, the airplane may be abandoned, ditched, or brought in for a dead-stick landing. To land with the engine dead, follow these instructions:

1. Lower nose at once so that airspeed does not drop below stalling speed. Keep IAS well above stalling speed.

2. If external tanks or bombs are installed, release them over an uncongested area, by pulling bomb-tank salvo levers.

3. Turn OFF fuel shutoff lever. Battery-disconnect switch OFF, except when electrical power is desired for lights or radio.

4. Choose an area for landing. If near a landing field, notify tower. Judge your turns carefully and plan to land into wind.

5. Duck head, lower seat, and release sliding canopy by pulling canopy emergency release handle.

6. If a long runway is available and time and altitude are sufficient to properly plan an approach, landing gear handle DN. If landing under any other condition, keep gear up; you stand less chance of injury by making a belly landing.

7. Wing flap lever approximately "30°," saving last 20 degrees of flap to overcome possible mistakes in judgment. Lower flaps fully when proper landing is ensured.

8. Land into wind, changing direction only as necessary to miss obstructions.

9. After landing, get out of airplane as quickly as possible and remain outside.

MAXIMUM GLIDE.

Maximum glide distance in event of a dead engine may be attained by gliding at an airspeed of 175 mph with gear and flaps up. If conditions permit, propeller control should be placed in full DECREASE in order to reduce drag as much as possible and to minimize windmilling. (See figure 3-2.)

FORCED LANDING (DEAD ENGINE).

See figure 3-3.

PRACTICE FORCED LANDING.

Forced landing can be simulated with the propeller control in full INCREASE to simulate the drag of a dead engine. Although optimum glide may be obtained with gear and flaps up, 10 degrees of flaps may be used to permit better visibility over the nose of the airplane without too seriously affecting the glide.

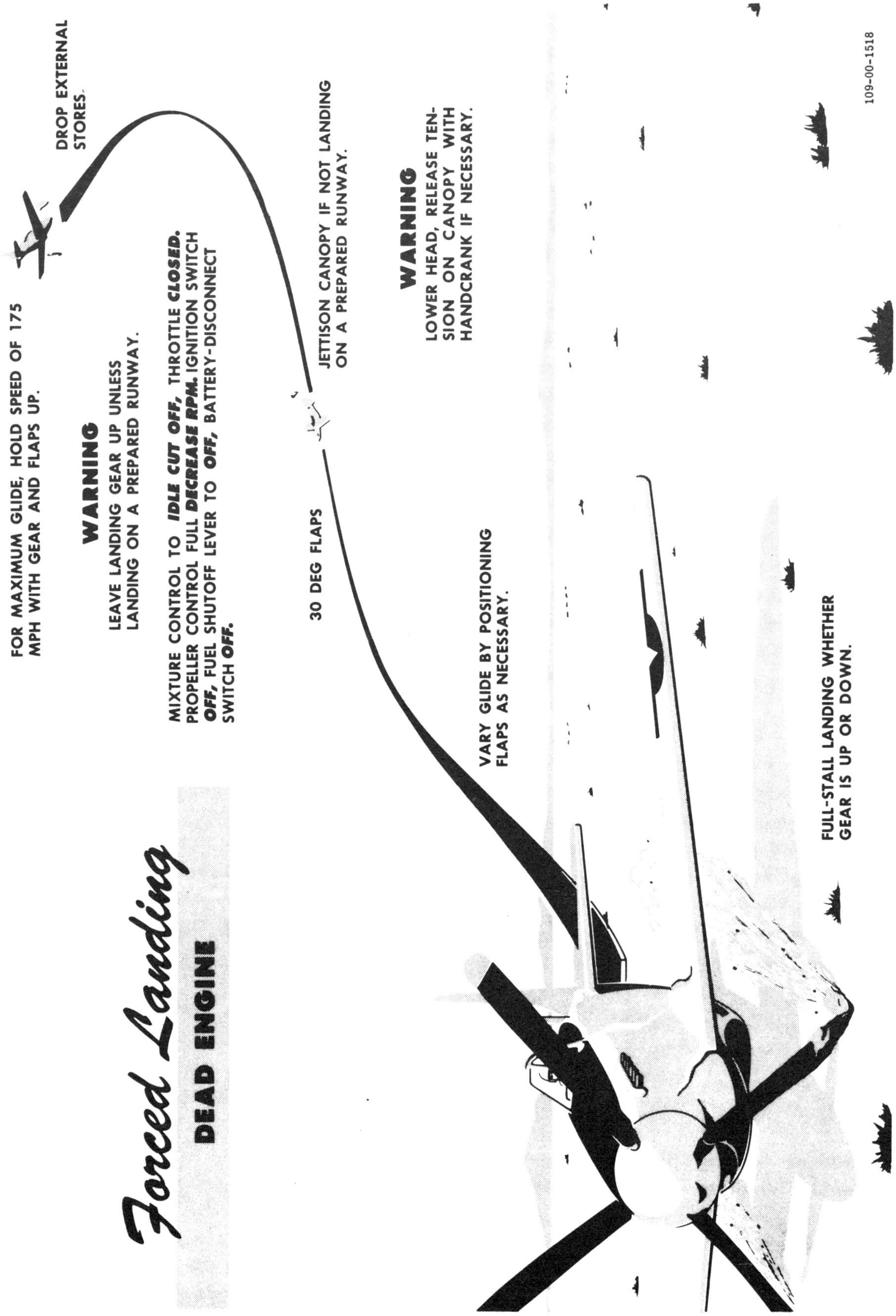
Forced Landing
DEAD ENGINE
FOR MAXIMUM GLIDE, HOLD SPEED OF 175 MPH WITH GEAR AND FLAPS UP.
WARNING
LEAVE LANDING GEAR UP UNLESS LANDING ON A PREPARED RUNWAY.
MIXTURE CONTROL TO IDLE CUT OFF, THROTTLE CLOSED. PROPELLER CONTROL FULL DECREASE RPM. IGNITION SWITCH OFF, FUEL SHUTOFF LEVER TO OFF, BATTERY-DISCONNECT SWITCH OFF.
DROP EXTERNAL STORES.
30 DEG FLAPS
JETTISON CANOPY IF NOT LANDING ON A PREPARED RUNWAY.
WARNING
LOWER HEAD, RELEASE TENSION ON CANOPY WITH HANDCRANK IF NECESSARY.
VARY GLIDE BY POSITIONING FLAPS AS NECESSARY.
FULL-STALL LANDING WHETHER GEAR IS UP OR DOWN.
109-00-1518

Figure 3-3

PROPELLER GOVERNOR FAILURE.

Failure of the governor to operate properly may result in a runaway propeller. A runaway propeller goes to full low pitch and may result in an engine rpm of 3600 or more. When such a failure occurs, the only method of reducing rpm is to pull the throttle back and decrease airspeed. In doing this, it is highly important to reduce the IAS to approximately 140 mph in order to obtain the maximum horsepower available. The following procedure is recommended:

1. Pull throttle back to obtain 3240 rpm.
2. Raise nose of airplane to lose speed, and then return to level-flight attitude. Keep IAS at approximately 140 mph.
3. When over landing area, lower gear and make approach at normal landing speed.

CAUTION

When engine speed and manifold pressure exceed allowable limits, the pilot should land at the nearest base and should record the duration of overspeed, the amount of overspeed, the manifold pressure, and (if known) the cause of overspeed.

FIRE.

Note

There is no fire-extinguishing system on this airplane.

ENGINE FIRE DURING STARTING.

If fire develops during starting, keep cranking engine in an attempt to blow fire out. If fire still persists:

1. Throttle CLOSED.
2. Mixture control IDLE CUTOFF.
3. Fuel shutoff lever OFF.
4. Battery-disconnect switch OFF.
5. Leave airplane as quickly as possible and signal ground crew to use portable fire-extinguishing system.

ENGINE FIRE AFTER STARTING.

If fire develops after starting, it will probably develop in carburetor area. Keep engine running for a short period, but if fire persists, follow procedure in preceding paragraph.

ENGINE FIRE DURING FLIGHT.

Depending upon the severity of the fire, either bail out immediately or shut down engine as follows:

1. Mixture control to IDLE CUTOFF.
2. Fuel shutoff lever OFF.
3. Throttle CLOSED.
4. Ignition switch OFF.
5. Battery-disconnect switch OFF except when power is desired to operate lights or radio.

FUSELAGE FIRE.

1. Reduce airspeed immediately in preparation for bail-out (if it becomes necessary) and to lessen possibility of fire spreading.
2. If smoke or fumes enter cockpit, use 100% oxygen and open canopy.
3. Generator-disconnect and battery-disconnect switches OFF.
4. If fire persists, shut down engine as outlined in preceding procedure.
5. If fire is not extinguished immediately, bail out.

WING FIRE.

If a wing fire develops, use the following procedure:

1. Turn off all wing light switches (position, identification, and landing), armament switches after jettisoning external load, and pitot heater switch.
2. Attempt to extinguish fire by sideslipping airplane away from flames.
3. If fire is not extinguished immediately, bail out.

ELECTRICAL FIRE.

Circuit breakers protect most electrical circuits and automatically interrupt power to prevent fire when a short occurs.

Note

Closing a circuit breaker that has opened in flight should be attempted only in case of emergency, and then only with full knowledge of the potential hazards involved and after careful evaluation of the advantages and the disadvantages.

If the defective circuit can be identified, the master switch for that circuit should be turned off. If fire still persists, turn battery-disconnect switch OFF. The generator-disconnect switch should be turned OFF if neither of the preceding is effective. Return to nearest available landing field as soon as possible, or, if fire tends to increase in intensity, bail out.

SMOKE ELIMINATION.

Should smoke or fumes enter the cockpit, proceed as follows:

1. Reduce airspeed immediately in preparation for bail-out and to minimize spreading of fire.
2. Open cold-air outlets.
3. Crank canopy to open position.
4. If smoke or fumes are still severe, use 100% oxygen as necessary.

Figure 3-4

LANDING EMERGENCIES.

BELLY LANDING.

If an emergency arises where a belly landing is indicated, proceed as follows:

1. Pull bomb-tank salvo levers to release external load.
2. Duck head and release sliding canopy by pulling canopy emergency release handle.
3. Just prior to contact, all switches off.
4. Lock shoulder harness.
5. Wing flap lever at approximately "30°," saving last 20 degrees to overcome possible mistakes in judgment.
6. Make normal flare-out and hold airplane off ground as long as possible.
7. After landing, get out of airplane as soon as possible.

EITHER GEAR UP OR UNLOCKED.

Ordinarily a wheels-up landing is preferable to a landing with only one wheel extended. However, if one wheel is extended and cannot be retracted, proceed as follows:

1. Jettison external load.
2. Roll canopy full back.
3. Lock shoulder harness.
4. Make normal flaps-down approach with wing low on extended-gear side.
5. Touch down on locked main wheel and tail wheel simultaneously, using ailerons to hold up wing with retracted or unlocked wheel.
6. Ignition switch OFF.
7. Maintain controlled ground roll by use of steerable tail wheel, brake, and rudder.
8. When wing tip strikes ground, apply maximum brake pressure possible to extended gear without nosing over.

EMERGENCY ENTRANCE.

An external canopy emergency release handle (figure 3-4) is located forward of the windshield bow on the upper right-hand longeron. Pulling the handle hard enough to break the light safety wire inside the cockpit releases the canopy so that it may be removed from the cockpit.

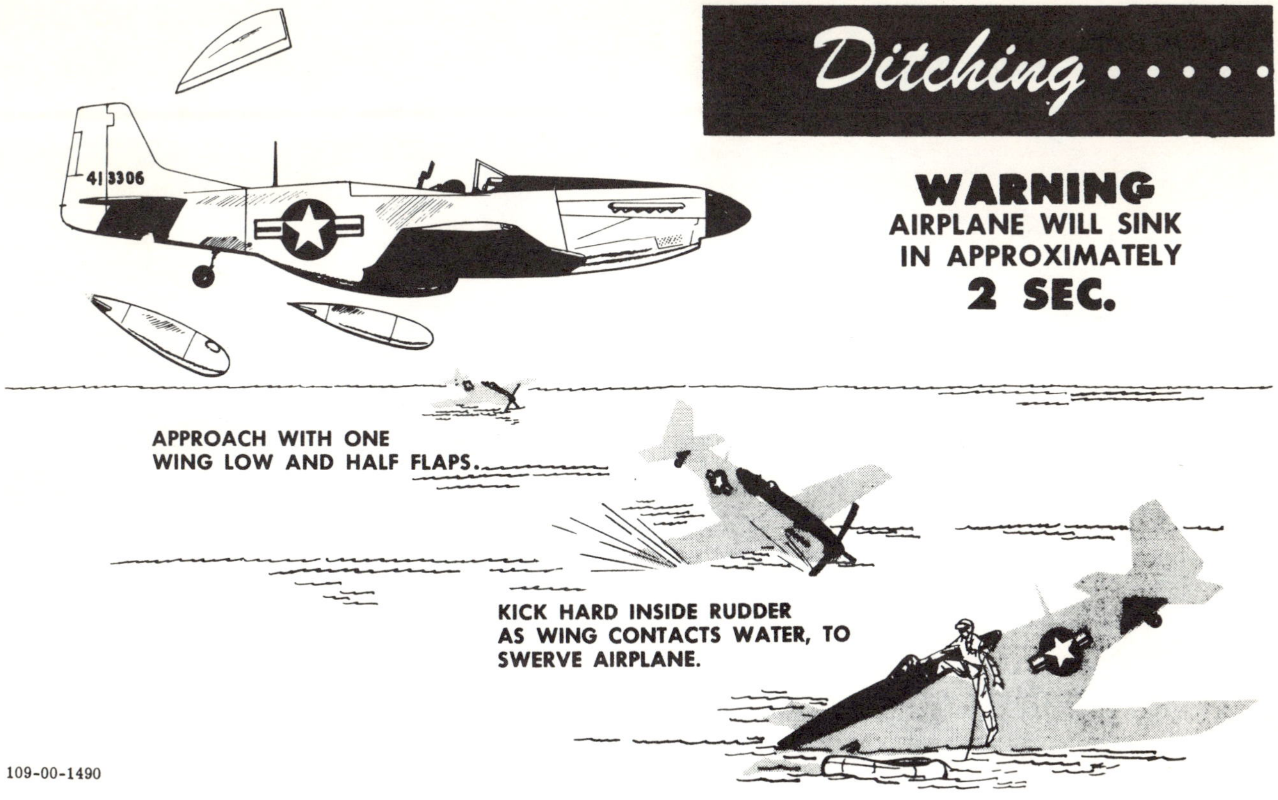

Figure 3-5

DITCHING.

The airplane should be ditched only as last resort. If it is impossible to maintain sufficient altitude for bail-out, ditch according to the following procedure (figure 3-5):

1. Follow radio distress procedure, giving location.
2. Jettison external load.
3. Unbuckle parachute; make sure life raft is fastened to you.
4. Lower seat and head as far as possible and pull canopy emergency release. Use crank if necessary to relieve pressure against windshield bow.
5. Tighten safety belt and lock shoulder harness because of high final impact.
6. Disconnect headset, oxygen equipment, and anti-G suit. Make sure no personal equipment will foul on your way out.
7. Check gear up and flaps one-half down.
8. Land into wind with one wing about 20 degrees low, and maintain enough speed above stall to keep rudder control. As low wing hits water, kick hard inside rudder to spin airplane around on surface to prevent severe diving and quick deceleration. As soon as airplane comes to rest, get out immediately.

BAIL-OUT.

When the decision is made to abandon the airplane and time permits, jettison external load (bombs, rockets, or tanks) if the area below is uninhabited. Before bail-out, reduce airspeed as much as possible and trim to slightly nose-down attitude. Head for an uninhabited area and follow procedure shown in figure 3-6.

ALTERNATE BAIL-OUT.

When airplane is controllable, the following bail-out procedure is recommended:

1. Disconnect radio, oxygen, and anti-G suit connections.
2. Release canopy after lowering seat and ducking head.
3. Roll airplane over on its back and trim for inverted climb.
4. Release safety belt and shoulder harness, and drop clear.

FUEL SYSTEM FAILURE.

If engine begins cutting out in flight and the fuel system is suspected, immediately change fuel tank selector handle. If condition persists, proceed as follows:

1. Release drop tanks if empty. (Air from empty drop tanks may leak past the fuel selector valve and permit the engine fuel pump to suck air.)

Bail-Out PROCEDURE

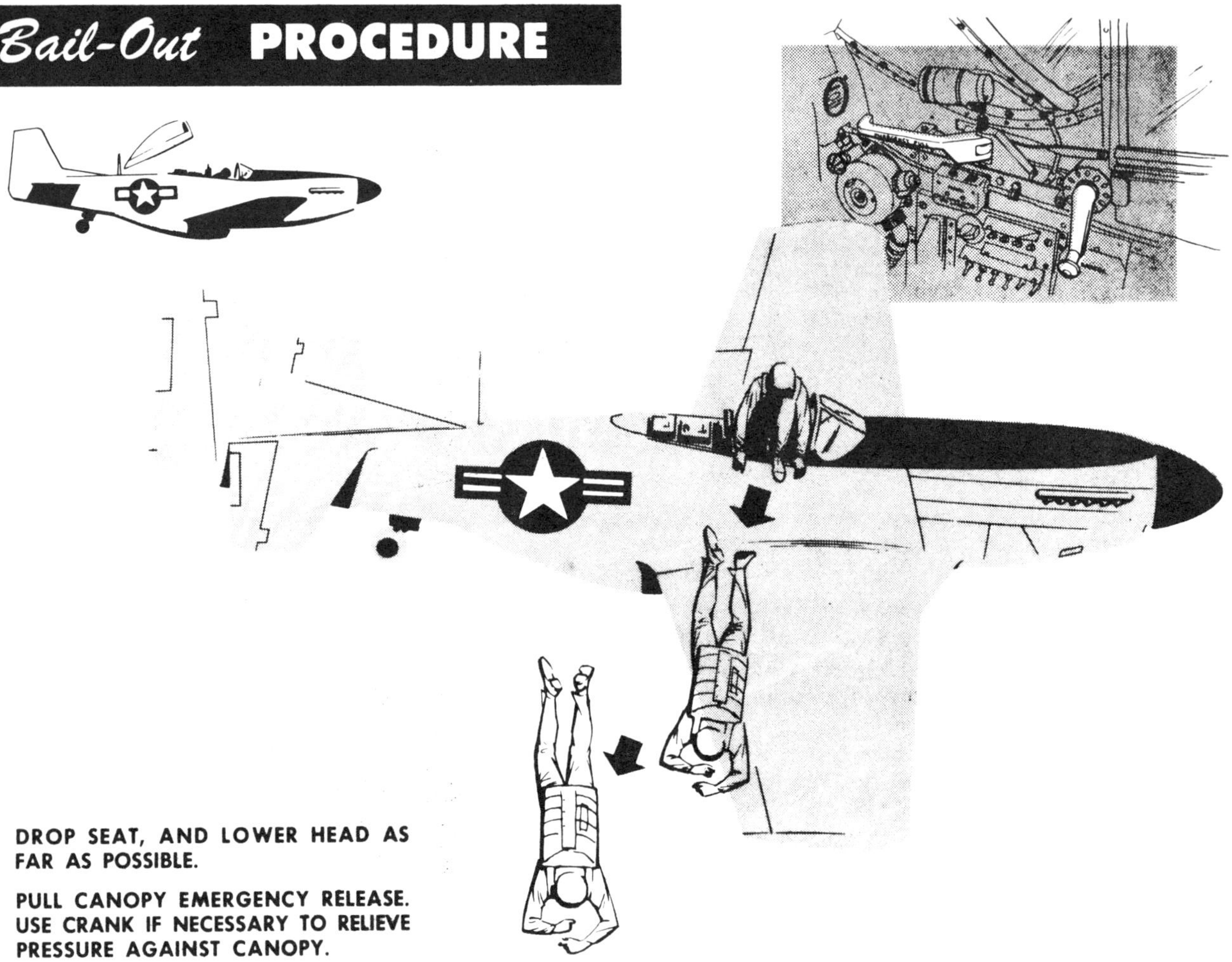

1 DROP SEAT, AND LOWER HEAD AS FAR AS POSSIBLE.

2 PULL CANOPY EMERGENCY RELEASE. USE CRANK IF NECESSARY TO RELIEVE PRESSURE AGAINST CANOPY.

3 JUST BEFORE LEAVING AIRPLANE, IF AT ALTITUDE, PULL BALL RELEASE KNOB ON BAIL-OUT BOTTLE. IF TIME PERMITS, DISCONNECT OXYGEN HOSE AND HEADSET, AND THEN RAISE SEAT TO TOPMOST ELEVATION.

4 CROUCH AS SHOWN AND DIVE TOWARD RIGHT WING TIP.

NOTE: RIGHT SIDE IS RECOMMENDED BECAUSE THE SLIP STREAM WILL HELP YOU CLEAR THE AIRPLANE. THE WING WILL THEN PASS YOUR BODY, OR IT WILL BE POSSIBLE TO SLIDE OFF THE WING WITHOUT STRIKING THE TAIL.

WARNING BAIL OUT ON OUTSIDE OF A SPIN TO MINIMIZE DANGER OF BEING STRUCK BY AIRPLANE.

109-00-1489

Figure 3-6

2. Reduce altitude to below 8500 feet. (Engine-driven fuel pump alone will supply fuel up to this altitude.)

3. If engine still cuts out after tanks are dropped, flight may possibly be continued at reduced power (1500 rpm) by use of the primer.

ELECTRICAL POWER SYSTEM FAILURE.

When a constant reading of more than 75 amperes is shown on the ammeter, either a very low-charged battery or a shorted circuit is indicated. Under these circumstances, leave the generator-disconnect switch ON, turn the battery-disconnect switch OFF, and check the system as follows:

1. If ammeter reading goes down to normal, it indicates a low battery. In this case, turn battery-disconnect switch ON again, checking, however, to see that ammeter reading goes down as battery builds up its charge.

2. If reading is still high and you are on the ground, return to ramp.

3. If short cannot be found, turn off all electrical circuits, including battery-disconnect and generator-disconnect switches. Use electrical system only when necessary, such as for checking and adjusting coolant temperatures.

4. Land at nearest available facility.

GENERATOR FAILURE.

If generator failure is suspected, the following method may be used for checking and continued operation:

1. Turn battery-disconnect switch OFF. If the electrical system continues to operate, the generator is functioning. If the electrical system operates and the ammeter shows no reading, the ammeter is faulty.

2. If electrical system fails to operate, generator is inoperative.

Note

In case of generator failure, care must be taken to conserve remaining energy of battery.

3. Turn ON battery-disconnect switch. Use battery power only when it is necessary to adjust coolant shutters or any other necessary electrical equipment.

4. Land at nearest available facility.

HYDRAULIC POWER SYSTEM FAILURE.

Hydraulic power system failure will affect the operation of the landing gear and wing flaps. No provision is made in the airplane for a pilot-actuated hydraulic hand-pump. If the engine-driven hydraulic pump fails and the rest of the system is intact, enough pressure will be supplied by the accumulator to lower flaps fully, provided the pressure gage shows at least 800 psi. The landing gear can be mechanically released from the up position in case of hydraulic failure.

LANDING GEAR EMERGENCY OPERATION.

LANDING GEAR EMERGENCY RETRACTION.

In event it is necessary to retract the landing gear during a landing or take-off run, the landing gear control handle should be moved to the UP position. The main gear will retract as long as the airplane is in motion. Main gear will not retract when the airplane is not in motion, even though the gear handle is placed in the UP position, as the hydraulic pressure does not have sufficient power to retract the landing gear while the airplane is stationary.

LANDING GEAR EMERGENCY LOWERING.

In event of hydraulic system failure, the landing gear may be lowered by placing the landing gear handle in the DN position and yawing the airplane. If red landing gear warning light illuminates or landing gear horn sounds when throttle is retarded, indicating an unsafe condition, pull fairing door emergency handle and yaw airplane to force gear into a locked position.

CAUTION

If landing gear handle requires excessive pressure to operate, do not force it into DN position, but release hydraulic pressure by pulling fairing door emergency release handle and then place landing gear handle in DN position.

CANOPY EMERGENCY OPERATION.

An emergency canopy release handle is located on the upper right longeron aft of the instrument panel. The handle is safetied with light-gage safety wire to prevent accidental operation. In emergency, jettison the canopy as follows:

1. Lower seat as far as possible.

2. Duck head and pull emergency canopy release handle.

WARNING

Be sure to lower the seat and to duck your head to avoid being hit by the canopy.

Note

If excessive force was used to secure canopy before take-off, it may be necessary to crank the canopy back to relieve pressure against the windshield before emergency release is effective.

DROP TANK EMERGENCY RELEASE.

To release bombs or droppable fuel tanks mechanically in an emergency, pull out both bomb and drop tank emergency release handles.

ENGINE OVERHEATING.

If the engine overheats in flight (indicated by coolant relief valve pop-off, maximum coolant temperature being exceeded, or white smoke coming from exhaust stacks), move coolant radiator air control switch to OPEN and hold. If, after approximately 20 to 30 seconds, the temperature still remains high, failure of the coolant flap actuator is indicated. Release coolant radiator air control switch and pull coolant flap emergency release handle. Reduce power to minimum necessary to maintain flight altitude. If overtemperature persists, land as soon as possible.

CAUTION

If conditions are favorable for a dead-stick landing and overtemperature persists, consider the possibility of shutting down the engine prior to landing.

If the high coolant temperature is not caused by actuator failure, an undesirable cooling condition may result from use of the coolant flap emergency release handle. To prevent excessive cooling, hold coolant radiator air control switch in CLOSE position for approximately 20 seconds after using the emergency release. This ensures that the flap is not extended beyond 7 inches if the electrical actuator is functioning at all. Then move switch to OFF for remainder of flight.

After the coolant flap emergency release handle is used, low-power engine operation should be avoided to prevent the coolant temperature from going below the minimum allowable as a result of the greater flap opening. There is no provision for emergency closing of the flap, nor can the emergency release be reset in flight.

CAUTION

Use the emergency release with discretion. High coolant temperatures may be the result of high power settings, low-altitude flight, engine malfunction, or a broken indicator, rather than actuator failure.

Auxiliary Equipment

DESCRIPTION AND OPERATION

SECTION IV

COCKPIT HEATING AND VENTILATING SYSTEM.

Warm air for heating the cockpit is routed from a scoop aft of the coolant radiator. (See figure 4-2.) The warm air passes through a flexible duct to a point behind the pilot's seat. From there, a duct leads to the cockpit hot-air outlet valve on the right side of the cockpit. A metal-spring pointer drags on a calibrated shoulder on the valve to prevent the gate from changing position because of vibration or force of airflow. Air from the forward section of the radiator air scoop is used to cool the cockpit.

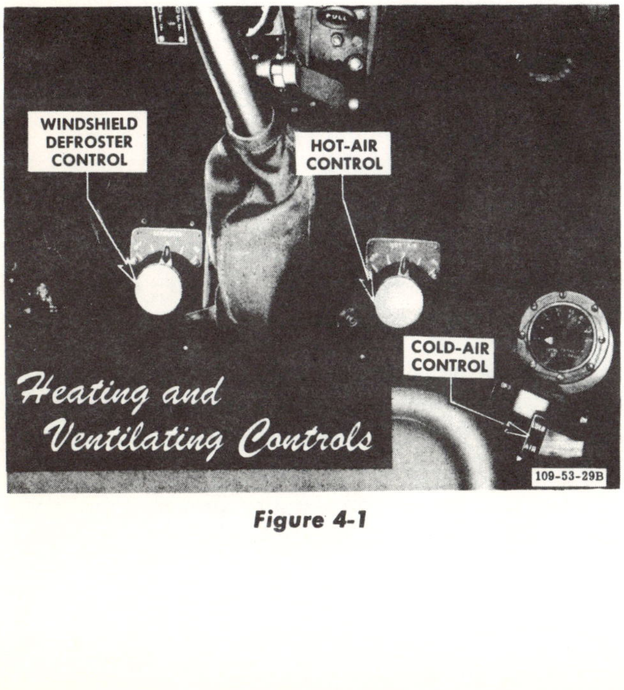

Figure 4-1

COCKPIT HEATING AND VENTILATING CONTROLS.

COCKPIT HOT-AIR CONTROL KNOB.

The cockpit hot-air control knob (figure 4-1) marked "HOT AIR," mechanically controls a gate-type valve and is located on the floor of the pilot's cockpit, below the right front edge of the seat. A pointer on the hot-air control knob may be positioned to ON, OFF, or any intermediate position.

COLD-AIR CONTROL HANDLE.

The cold-air control handle (figure 4-1), located at the right side of pilot's seat on the cockpit floor, has ON and OFF positions. Operation of the control handle mechanically allows a flow of cold air from the two outlets located behind the pilot's seat.

DEFROSTING SYSTEM.

The hot air for windshield defrosting is obtained from the same port aft of the cooling radiator as the cockpit warm air. The windshield defrost air flows through a flexible duct to the forward part of the cabin, where it is diverted into three separate ducts and conducted to the front windshield and side glass panels.

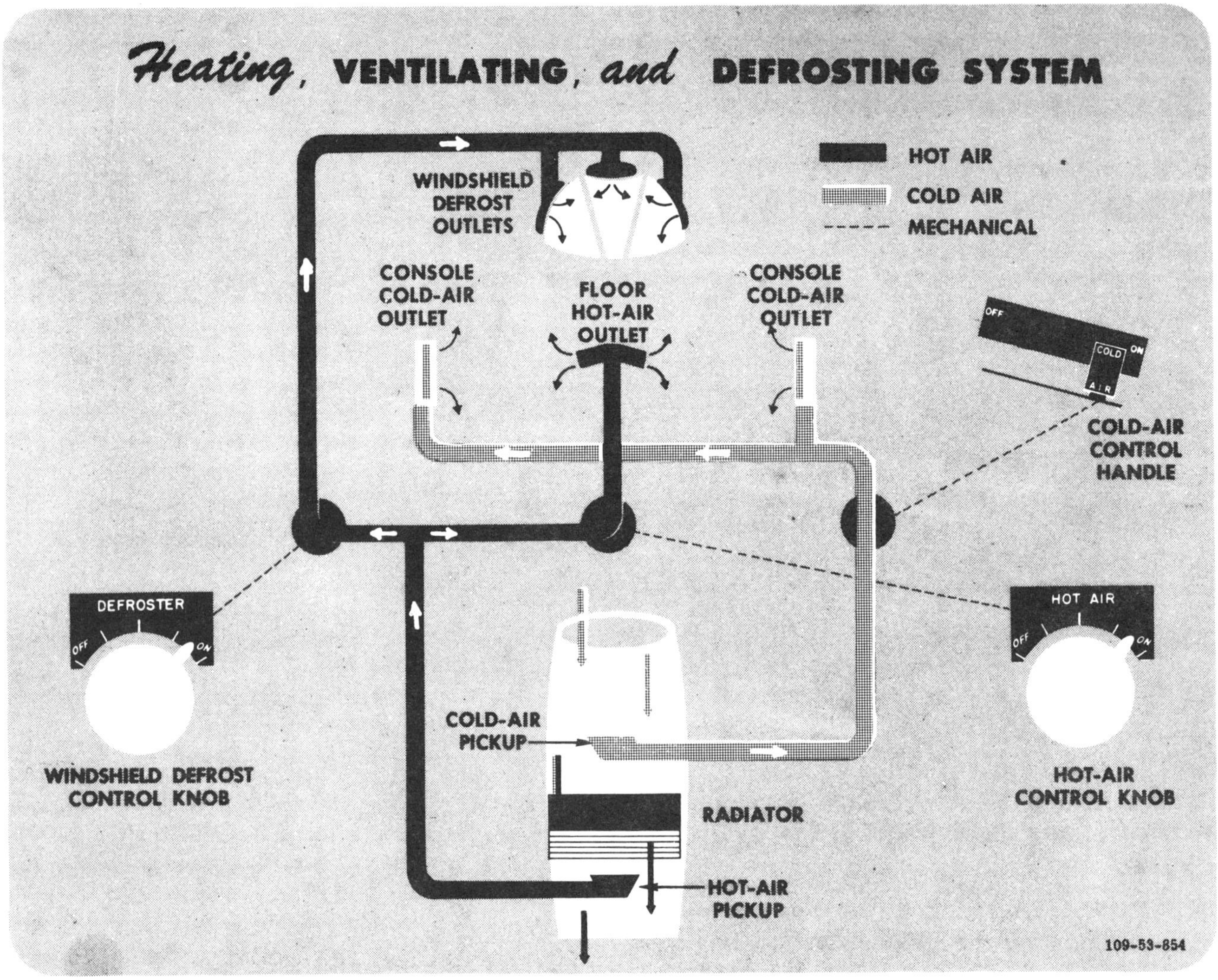

Figure 4-2

DEFROSTING CONTROL KNOB.

The windshield defrosting control knob (figure 4-1) is located on the floor of the cockpit, on the left side below the seat. The control may be positioned at ON, OFF, or any intermediate position and mechanically controls a gate valve in the air duct.

PITOT HEATER.

The pitot head, located on the underside of the right wing, is equipped with two resistance-type heaters in the forward end. The heaters prevent the formation of ice and, consequently, prevent erroneous airspeed indicator readings.

PITOT HEATER SWITCH.

A pitot heater switch (figure 4-6), located on the right switch panel, has ON and OFF positions.

CAUTION

To prevent unnecessary discharging of the battery, the pitot heater switch must be OFF when the airplane is on the ground or when heat is not required. Prolonged ground operation will permanently damage the unit, as there is insufficient cooling.

COMMUNICATION AND ASSOCIATED ELECTRONIC EQUIPMENT.

Radio sets installed in the airplane are the SCR-522-A or AN/ARC-3 command set, the BC-453-B range receiver installed in conjunction with the command set, the SCR-695-A IFF set, and the AN/ARA-8 homing adapter. (See figure 4-3.) On airplanes equipped with fuselage fuel tank and battery behind the pilot's seat,

Communication
AND ASSOCIATED ELECTRONIC EQUIPMENT

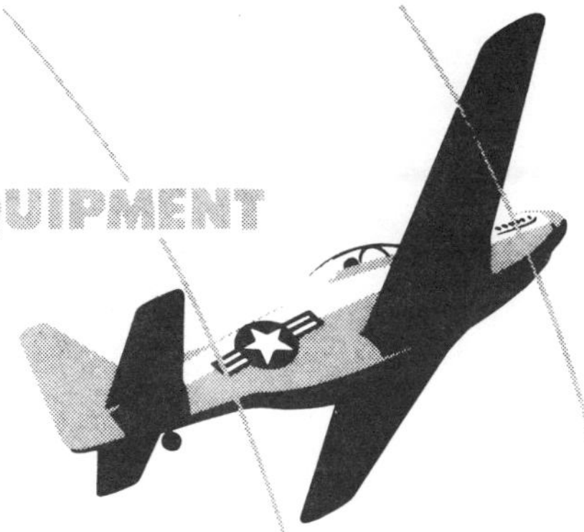

TYPE	DESIG-NATION	USE	RANGE	LOCATION OF CONTROLS
COMMAND	AN/ARC-3 OR SCR-522A	TWO-WAY COMMUNICATION	30 MI AT 1000 FT, 135 MI AT 10,000 FT	RIGHT SIDE OF COCKPIT
RADIO RANGE RECEIVER	BC-453B	RADIO RANGE RECEPTION	50 TO 70 MILES	RIGHT SIDE OF COCKPIT
HOMING RECEIVER	AN/ARA-8	HOMING	30 MI AT 1000 FT, 135 MI AT 10,000 FT	RIGHT SIDE OF COCKPIT
IFF	SCR-695A OR AN/APX-6	IDENTIFICATION	LINE OF SIGHT	RIGHT SIDE OF COCKPIT

109-71-907

Figure 4-3

the IFF set is not installed. However, if either the fuselage fuel tank has been removed or the battery has been moved forward of the fire wall, the IFF equipment may be installed. For antenna installations, see figure 4-5.

SCR-522-A COMMAND RADIO.

This set is a push-button controlled transmitter-receiver that operates on the 100 to 156 mc band. The control box is just aft of the right switch panel in the cockpit. A microphone button is located on the throttle lever. On some airplanes, a remote contactor is installed on the left side of the instrument panel. The contactor switches the transmitter from the "A," "B," or "C" band to the "D" band for 14 seconds of every minute. The pointer on the face of the contactor indicates when the switching action takes place. Normally, the clock switch on the contactor should not be touched in flight; it is set on the ground by the service crew.

OPERATION OF SCR-522-A COMMAND RADIO.

Note

> The "T-R-REM." switch is lockwired in the REM position.

To receive or transmit on channel "A," "B," "C," or "D," press corresponding channel selector button on control box. Tubes require approximately 30 seconds to warm up. Adjust headset volume with volume control on junction box, and monitor station to be contacted. On airplanes equipped with a remote contactor, check operation with switch in OUT and IN positions. Press throttle microphone button and speak in a normal tone. To receive, release pressure on microphone button.

Note

> Indicator lamp brilliance is controlled by the dimmer mask lever on the control box. The lamps behind the four green jewels indicate

Figure 4-4

the channel in operation. The lamp behind the white jewel opposite the "T-R-REM." switch glows when the equipment is in the receive condition.

To turn set off, press "OFF" button on control box.

AN/ARC-3 COMMAND RADIO.

The AN/ARC-3 set consists of a receiver-transmitter, a power supply, and a control box. The receiver-transmitter is installed on a shelf behind the pilot's seat, the power supply is on the floor behind the pilot's seat, and the control box is located on the radio control panel at the right side of cockpit. The set provides two-way communication from airplane to airplane or from airplane to ground within a frequency range of 100 to 156 megacycles. Eight preset channels are provided, and the set has a line-of-sight range. Average range is approximately 30 miles at an altitude of 1000 feet and 135 miles at 10,000 feet.

OPERATION OF AN/ARC-3 COMMAND RADIO.

To operate the AN/ARC-3 command radio, proceed as follows:

1. Push desired channel button (figure 4-4) on control box and allow approximately 30 seconds for set to warm up. When audio tone that is heard in headset during latter portion of warm-up period stops, set is ready for operation.

2. Push channel selector button to change frequency channel.

3. Adjust volume with volume control knob, located directly above control box.

4. To transmit, press microphone button on throttle control and use microphone.

5. To restore reception, release microphone button.

6. To turn off set, press off button on control box.

CAUTION

Do not attempt to select another channel until tuning cycle is completed. The tuning mechanism is motor-driven, and, once a selector button is pressed, the tuning mechanism is set in motion. When the set is tuned to the selected frequency, the tuning drive mechanism automatically stops. At this time, an electrical relay arrangement is established, permitting the drive mechanism and drive motor to start another tuning cycle when another range selector button is pressed. If, however, a tuning cycle is interrupted before completion, the relay and mechanism are not in proper condition to permit further tuning. During a tuning cycle, a constant audio note may be heard in the headphones. When this note ceases, it indicates completion of the tuning cycle. The control may then be tuned through another cycle to a different frequency. If the set is turned off by the off button, do not attempt to turn it on again immediately, but wait at least one minute before doing so. Failure to observe these precautions may result in the band-shifting mechanism becoming inoperative with the set not tuned to any frequency. If this happens, the equipment must be readjusted in order to operate. Attempting to select another channel while the engine is turning less than 1800 rpm may have the same effect on the set.

BC-453-B RANGE RECEIVER.

The BC-453-B range receiver is mounted on a shelf in the upper aft part of the fuselage. The receiver operates on frequencies between 0.19 and 0.55 mc and is primarily used for reception of weather reports, beacon signals, and airport communications. This receiver is used in conjunction with the AN/ARC-3 radio.

OPERATION OF BC-453-B RANGE RECEIVER.

Operate the BC-453-B receiver as follows:

1. Place selector switch (figure 4-4) in MCW position.
2. Move range toggle switch to RANGE position.
3. Rotate left selector switch to NORMAL USE.
4. Rotate right selector switch to MIX SIGNALS AND COMMAND.
5. Turn up volume by rotating "VOLUME" knob as desired.
6. Rotate tuning crank to tune desired signal.

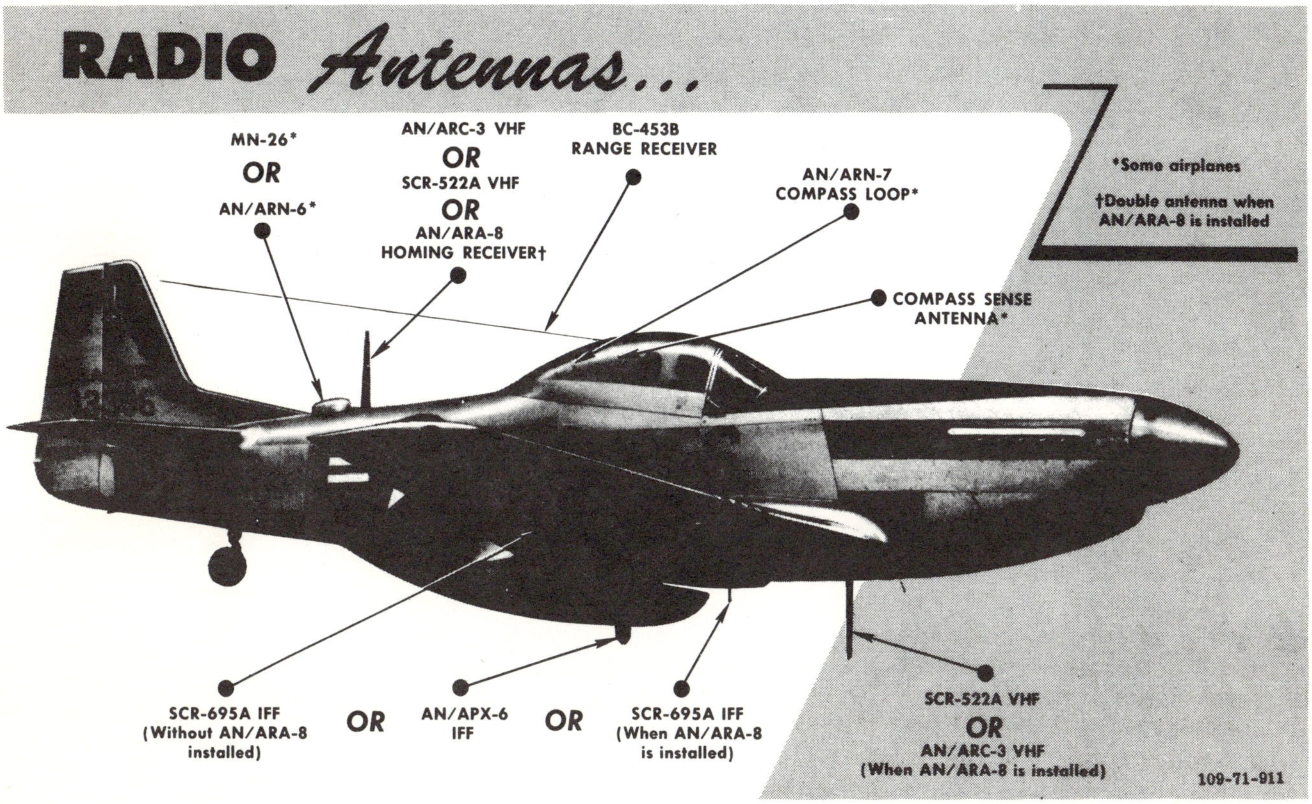

Figure 4-5

AN/ARA-8 HOMING ADAPTER.

This is an adapter unit used in conjunction with the AN/ARC-3 vhf radio to permit homing on any transmitted carrier within the frequency range of 120 to 140 megacycles. In addition, this equipment may be used for air-to-air homing for purposes of rendezvous. Homing can be accomplished on cw, mcw, and audio pulse signals. Controls are provided above the vhf control box at the right side of the cockpit.

OPERATION OF AN/ARA-8 HOMING ADAPTER.

1. To operate homing adapter, move "HOMING-COMM.-TRANS." switch (figure 4-4) to HOMING position.

2. To stop operation of homing adapter, move "HOMING-COMM.-TRANS." switch to COMM. position.

SCR-695-A IDENTIFICATION RADIO.

The SCR-695-A radio set permits automatic transmission of identification signals upon reception of a challenging signal from properly equipped friendly air or surface units. It can also be used to transmit emergency or distress signals. The IFF controls include a code selector which provides a choice of six code settings, an emergency switch for transmitting a distress signal, and an on-off control switch. The set can be operated from sea level to approximately 50,000 feet. Destructor units have been removed from this equipment.

OPERATION OF SCR-695-A IDENTIFICATION RADIO.

The SCR-695-A IFF set is operated as follows:

1. Rotate code selector (figure 4-4) to desired position.

2. Move on-off switch to ON position.

3. Move on-off time switch to ON position.

4. If emergency or distress signal is needed, lift guarded switch to ON position.

5. To turn set off, move on-off switch to OFF position.

6. Move on-off time switch to OFF position.

AN/APX-6 IDENTIFICATION RADAR.

The AN/APX-6 radar identification set (if installed) is used to automatically identify the airplane as friendly whenever it is properly challenged by suitably equipped friendly air or surface forces. The set also has provision for identifying specific friendly airplanes within a group and means for transmitting a special distress code. Functionally, the AN/APX-6 set receives challenges and transmits replies to the source of the challenges, where the replies are displayed, together with the associated radar targets, on radar indicators. When a radar target is accompanied by a proper reply from the IFF set, the target is considered friendly. Three destructors, mounted in the AN/APX-6 transpondor, may be actuated by the pilot. An impact switch automatically actuates the destructors upon a crash landing.

OPERATION OF AN/APX-6 IDENTIFICATION RADAR.

CAUTION

Before take-off, check that AN/APX-6 frequency counters have been set to proper frequency channels. (The IFF units are accessible behind the pilot's armor plate under the canopy.)

1. To turn equipment on, rotate master control to NORM position (full sensitivity and maximum performance).

2. Rotate master control to STDBY to maintain equipment inoperative but ready for instant use.

Note

The LOW position of the master control (partial sensitivity) should not be used except upon proper authorization.

3. Set three-position "MODE 2" and "MODE 3" switches to their OUT and I/P positions unless otherwise directed.

4. For emergency operation, press dial stop and rotate master control to EMERGENCY position so that set will automatically transmit distress signals.

5. To manually fire destructors, lift guard and move destructor switch to ON.

6. To turn off IFF set, rotate master control to OFF.

7. If AN/APX-6 transpondor is destroyed during flight, report this information immediately after landing.

CAUTION

The destructors should be fired only when the AN/APX-6 equipment is in danger of falling into enemy hands. If a forced landing has to be made in an area of doubtful security, fire destructors.

LIGHTING EQUIPMENT.

EXTERIOR LIGHTING.

Exterior lighting consists of a landing light and position lights. The automatically retracting landing light is located in the left main landing gear well. Type A-8 position lights are located on the outer tip of each wing and on the trailing edge of the rudder. The position lights on the left wing tip are red; on the right wing tip, green; and on the rudder, white.

CAUTION

While the airplane is on the ground, do not allow the lights to burn for any prolonged length of time, as the heat will seriously damage the lenses.

EXTERIOR LIGHTING CONTROLS.

LANDING LIGHT SWITCH. The landing light switch (8, figure 1-4), located on the left switch panel aft of the throttle quadrant. The switch has ON and OFF positions to illuminate the light only. Retraction and extension of the light is controlled automatically by mechanical means when the gear is operated. A safety switch is incorporated into the landing light circuit to cut off current to the landing light when the light is retracted.

CAUTION

Keep ground operation of landing light to a minimum to prevent overheating and damaging unit.

POSITION LIGHT SWITCHES. Two separate switches and circuits are used to operate the position lights. The switches (figure 4-6) are located on the right switch panel and are marked "WING" and "TAIL." Each switch has two positions, DIM and BRIGHT. Current for the left wing position light passes through a pair of twisted wires to reduce magnetic influence on the remote compass.

INTERIOR LIGHTING.

Instrument lighting is provided by two fluorescent lights, one on each side of the instrument panel. A rotating light hood on each light controls the area lighted, and a separate rheostat controls intensity. In addition, a cockpit light is installed on each side of the cockpit. On early airplanes, provision is made for stowing a signal lamp on the left side of the cockpit floor. Colored filters may be used with the lamp.

Figure 4-6

INTERIOR LIGHTING CONTROLS.

INSTRUMENT PANEL LIGHT RHEOSTATS. The right and left fluorescent instrument panel lights have individual rheostats located on the right switch panel (5, figure 1-5) and on the radiator air control panel (9, figure 1-4), respectively. The rheostats operate alike and must be turned clockwise to the START position. When the light goes on, the rheostat may be turned to DIM or ON as desired.

COCKPIT LIGHT RHEOSTAT. A cockpit light is located on each side of the cockpit. Each light has a switch built into the light assembly. A single rheostat (28, figure 1-6), located on the front switch panel, controls the intensity of both lights.

OXYGEN SYSTEM.

Oxygen is supplied from two Type D-2 and two Type F-2 low-pressure oxygen cylinders located in the aft fuselage. A blinker-type flow indicator operates with the breathing of the pilot, indicating proper operation of the system. The oxygen cylinders may be refilled without removal from the airplane, by means of a filler valve (figure 1-25) located on the lower left side of the fuselage. Normal full pressure of the oxygen system is 400 psi. For oxygen duration at different altitudes, refer to oxygen duration chart (figure 4-7).

Note

As an airplane ascends to high altitudes, where the temperature is normally quite low, the oxygen cylinders become chilled. As the cylinders grow colder, the oxygen gage pressure is reduced, sometimes rather rapidly. With a 100°F decrease in temperature in the cylinders, the gage pressure can be expected to drop 20

Oxygen Duration —HOURS

COCKPIT ALTITUDE —FEET—	GAGE PRESSURE—PSI 400	350	300	250	200	150	100	BELOW 100
40,000	8.5	7.3	6.1	4.9	3.6	2.4	1.2	EMERGENCY—DESCEND TO ALTITUDE NOT REQUIRING OXYGEN
	8.5	7.3	6.1	4.9	3.6	2.4	1.2	
35,000	8.5	7.3	6.1	4.9	3.6	2.4	1.2	
	8.5	7.3	6.1	4.9	3.6	2.4	1.2	
30,000	6.3	5.4	4.5	3.6	2.7	1.8	0.9	
	6.3	5.4	4.5	3.6	2.7	1.8	0.9	
25,000	5.0	4.3	3.6	2.9	2.2	1.4	0.7	
	6.0	5.1	4.3	3.4	2.6	1.7	0.9	
20,000	4.1	3.5	2.9	2.3	1.7	1.2	0.6	
	6.7	5.8	4.8	3.9	2.9	1.9	1.0	
15,000	3.2	2.7	2.3	1.8	1.4	0.9	0.5	
	8.2	7.0	5.8	4.7	3.5	2.3	1.2	
10,000	2.7	2.3	2.0	1.6	1.2	0.8	0.4	
	10.8	9.3	7.7	6.2	4.6	3.1	1.5	

BLACK FIGURES INDICATE DILUTER LEVER ***NORMAL OXYGEN***
TOP FIGURES INDICATE DILUTER LEVER ***100% OXYGEN***
CYLINDERS: TWO TYPE D-2, TWO TYPE F-2

Figure 4-7

percent. This rapid fall in pressure is occasionally a cause for unnecessary alarm. All the oxygen is still there, and as the airplane descends to warmer altitudes, the pressure will tend to rise again, so the rate of oxygen usage may appear to be slower than normal. A rapid fall in oxygen pressure while the airplane is in level flight, or while it is descending, is not ordinarily due to falling temperature, of course. When this happens, leakage or loss of oxygen must be suspected.

OXYGEN REGULATOR.

Early airplanes use the AN6004 oxygen regulator, while later airplanes use the A-12 diluter-demand regulator (1, figure 1-5; figure 4-8). The regulator is installed on the right side of the cockpit, just aft of the instrument panel. The AN6004 regulator has a diaphragm which actuates a valve, permitting oxygen to flow through the regulator, where it mixes with free air in varying amounts in accordance with barometric pressure. A control allows the user to close the air intake, thus causing pure oxygen to flow to the mask. The regulator also has an emergency valve, which causes oxygen to by-pass the regulator and flow directly to the mask. A take-off line directs oxygen into the blinker flow indicator to show when the regulator is functioning. On later airplanes, the Type A-12 diluter-demand regulator is installed. This regulator operates as a demand regulator to an altitude of 32,000 feet. Above 32,000 feet, the control knob may be turned to cause the regulator to deliver pure oxygen under pressure, regardless of the position of the diluter lever.

OXYGEN REGULATOR CONTROLS.

DILUTER LEVER (AN6004 REGULATOR). On the AN6004 regulator, a diluter lever is located on the side of the regulator case and may be positioned to NORMAL OXYGEN or 100% OXYGEN. The diluter lever manually operates the air shutoff valve, allowing the regulator to deliver pure oxygen when the lever is in the 100% OXYGEN position. During normal usage, the diluter lever is left in the NORMAL OXYGEN position to allow air and oxygen to mix in the proper proportions for any given altitude.

EMERGENCY VALVE (AN6004 REGULATOR). The regulator has an independent oxygen emergency valve. When the valve is turned counterclockwise, a continuous stream of oxygen is allowed to by-pass the regulator and flow to the mask.

DILUTER LEVER (A-12 REGULATOR). The diluter lever on the A-12 regulator is marked "AUTO. MIX." and has ON and OFF positions. With the diluter lever at

ON, oxygen is diluted with air of the proper proportions for any given altitude and should be ON for all normal operation. With the diluter lever at OFF, the regulator delivers pure oxygen to the mask.

EMERGENCY KNOB (A-12 REGULATOR). An emergency knob marked "EMER ON" has an arrow indicating the direction to turn the knob in event of failure of the regulator mechanism. The emergency knob allows pure oxygen to flow to the mask, by-passing the regulator mechanism.

OXYGEN REGULATOR INDICATORS.

PRESSURE GAGE. An oxygen pressure gage (19, figure 1-6) is located below the lower right side of the instrument panel. The gage is calibrated to read as high as 500 psi and has a red arc in the 0 to 100 psi range. For normal operation, the gage should show a full reading of 400 psi to 100 psi.

OXYGEN PRESSURE WARNING LIGHT (EARLY AIRPLANES). An oxygen pressure warning light (20, figure 1-6), located to the right of the oxygen pressure gage, illuminates to show when oxygen pressure drops to an unsafe value.

OXYGEN FLOW INDICATOR. On early airplanes, an oxygen flow indicator (14, figure 1-6) is located on the right side of the instrument panel; on later airplanes, the indicator is located to the right of the oxygen pressure gage below the right side of the instrument panel. A bellows assembly, actuated by the pilot's breathing, opens and closes a shutter on the face of the indicator to indicate normal oxygen system operation. Oxygen duration is shown in figure 4-7.

OXYGEN SYSTEM PREFLIGHT CHECK.

Before each flight requiring the use of oxygen, check the system as follows:

1. Check oxygen pressure gage for indication between 400 and 450 psi if night flight or flight above 10,000 feet is planned. If it is definitely known that a maximum flight altitude of 10,000 feet will not be exceeded or night flying is not contemplated, the pressure in the oxygen system must be at least 100 psi prior to flight. Should any doubt exist, however, as to adverse weather conditions that may be encountered on a long-range flight, the oxygen system must be charged to full capacity before take-off.

2. Check regulator diaphragm for leakage, with diluter lever set at NORMAL OXYGEN (ON if A-12 regulator), by placing open end of mask-to-regulator tube against mouth and blowing lightly into it for about 5 seconds. Any escape of air from the regulator indicates either a leaky diaphragm or a faulty check valve in the air inlet, and the regulator must be replaced before flight.

3. Check oxygen mask for fit and for absence of leakage.

4. Connect mask tube to regulator outlet. Check connection for tightness. Attach tube clip to parachute harness high enough to permit free movement of head without pinching or pulling face.

5. Breathe normally several times with diluter lever at NORMAL OXYGEN (ON if A-12 regulator) and then at 100% OXYGEN to check flow from oxygen regulator and operation of flow indicator.

6. Check oxygen regulator to see that emergency valve is safety-wired closed with light wire and that diluter lever is at NORMAL OXYGEN (ON if A-12 regulator).

ARMAMENT EQUIPMENT.

Armament provisions include machine guns, a gun sight, bombing and rocket equipment, and a camera. The armament control switches are located on a panel (figure 4-10) at the lower left side of the instrument panel. On airplanes with zero-rail rocket installation, the armament control switches are on the front switch panel (figure 4-9) and most of the engine control switches are on a panel (figure 1-15) at the lower left

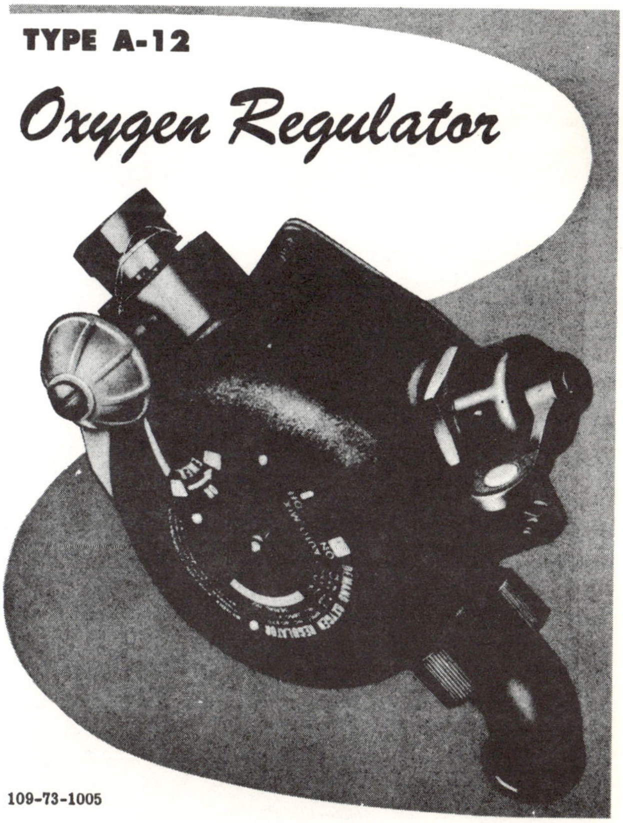

Figure 4-8

Figure 4-9

side of the instrument panel. The armament equipment derives its electrical power from the 28-volt direct-current system.

GUNNERY EQUIPMENT.

Gunnery equipment consists of complete provision for installation and operation of six fixed .50-caliber machine guns mounted in the wings. Each gun is provided with a Type J-1 or J-4 electric gun heater. Either of two gun installations is possible: (1) three fixed .50-caliber guns may be installed in each wing, with 500 rounds (400 rounds in late airplanes) of ammunition for each inboard gun and 270 rounds for each center and outboard gun; (2) the center gun may be removed, allowing the inboard guns to carry 500 rounds (400 rounds in late airplanes) each and the outboard guns to carry 500 rounds each. Ammunition containers are mounted in the wings; empty cases are ejected through the bottom of the wing. Gun charging is manually accomplished on the ground before flight. The guns are normally bore sighted with a point of convergence at 250 or 300 yards. A GSAP gun camera, installed in the leading edge of the left wing, inboard of the guns, is operated automatically when the guns or rockets are fired and may be operated independently.

GUNNERY EQUIPMENT CONTROLS.

GUN SAFETY SWITCH. The gun safety switch (figure 4-12), located on the right of the gun sight, has three positions: OFF, GUN, and CAMERA AND SIGHT. On combat missions, the switch should be moved to GUN as soon as the airplane is safely off the ground. With the gun safety switch at GUN, pressing the trigger all the way back fires the guns and operates the camera simultaneously. When use of the sight and camera only

Figure 4-10

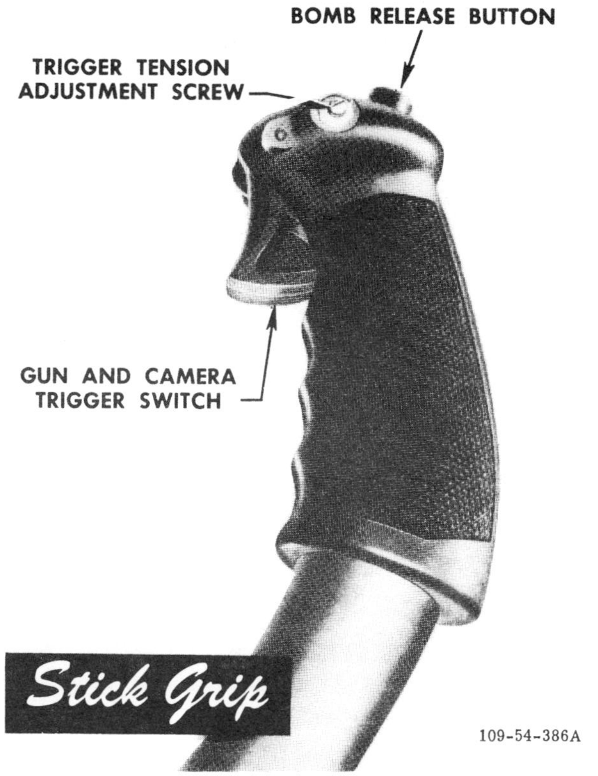

Figure 4-11

is desired, the switch should be positioned at CAMERA AND SIGHT.

TRIGGER. The gun trigger, located on the control stick grip (figure 4-11), has two positions. With the gun safety switch at GUN, pressing the trigger to the first position operates the camera only. Pressing the trigger still farther to the second position fires the guns and operates the camera simultaneously. On some early airplanes, the trigger is wired so that gun and camera operation is the same at both trigger positions. On these airplanes, only the position of the gun safety switch determines whether guns and camera, or camera alone operates. A placard on the armament switch panel of these airplanes reads "CAUTION—FIRST TRIGGER POSITION FIRES GUN AND CAMERA."

GUN HEATER SWITCH. The gun heater switch (figure 4-6) is located on the right switch panel. On missions that require gun heat, the gun heater switch must be moved to the ON position immediately after the engine is started. The switch should be returned to the OFF position before landing upon completion of the mission.

GUN CHARGER HANDLE. A gun charger handle is stowed in each gun bay for manually charging the guns before flight.

K-14A OR K-14B COMPUTING GUN SIGHT.

The K-14A or K-14B gun sight (figure 4-12) on late airplanes computes the correct lead angle for target crossing speed at ranges of from 200 to 800 yards. The sight contains two optical systems, fixed and gyro. The fixed optical system projects on the reflector glass a cross surrounded by a 70-mil ring. The 70-mil ring can be blanked out by the reticle masking lever on the left of the sight. Normally blanked out, the ring is used only in case of mechanical failure of the gyro or for ground strafing. The gyro optical system projects on the reflector glass a circle of six diamonds surrounding a central dot. The diameter of the circle is varied by changing the setting of the span scale lever on the face of the sight or by rotating the throttle twist grip.

K-14A OR K-14B GUN SIGHT CONTROLS.

GYRO MOTOR SWITCH. When the K-14A sight is installed, the sight gyro motor is turned on by a switch located on the selector-dimmer panel. Switch positions are ON and OFF. When the K-14B sight is used, this switch is not installed and the sight gyro is turned on when the battery-disconnect switch is moved to the ON position.

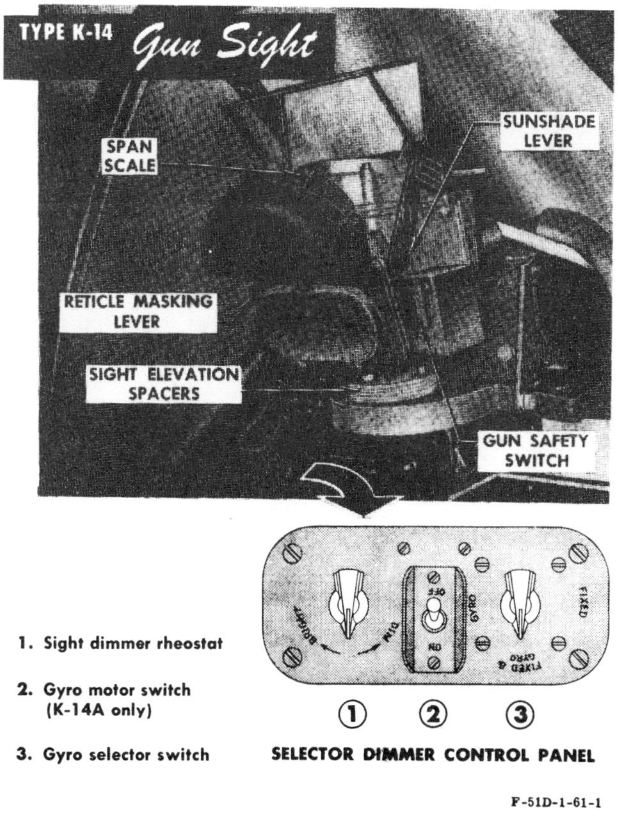

Figure 4-12

SIGHT DIMMER RHEOSTAT. On early airplanes, the sight dimmer rheostat (figure 1-6) is located on the selector-dimmer control panel, below the left side of the instrument shroud, and has BRIGHT and DIM positions. On late airplanes, the sight dimmer rheostat is on the front switch panel (figure 1-15), below the instrument panel, and has OFF and ON positions. The rheostat can be set at intermediate positions for varying degrees of brilliance.

GYRO SELECTOR SWITCH. The sight gyro is controlled by a selector switch (figure 4-12) with FIXED, FIXED & GYRO, and GYRO positions. The three positions allow the sight to be used as a fixed sight, combined fixed and compensating sight, or compensating sight only. During landing, the switch should be at FIXED to prevent damage to gyro. The gyro selector switch is located on the selector-dimmer panel below the left instrument shroud.

THROTTLE TWIST GRIP. The twist grip on the throttle (figure 1-8) of late airplanes is used to adjust sight range. Turning the twist grip clockwise enlarges the target circle on the reticle; turning it counterclockwise reduces the circle. When the target is properly placed in the circle, the sight computes the correct lead for the range.

PREFLIGHT CHECK OF K-14A OR K-14B GUN SIGHT.

Before take-off, check the sight as follows:

1. Gun safety switch at CAMERA AND SIGHT.
2. Gyro selector switch at FIXED & GYRO. Both reticle images should appear on reflector glass.
3. Rotate sight dimmer rheostat to obtain desired reticle brilliance.
4. Pick a point on horizon; make sure gyro reticle image dot is superimposed on fixed-reticle cross.
5. Rotate throttle twist grip to check operation of gyro reticle image circle from minimum to maximum range.

FIRING GUNS WITH K-14A OR K-14B GUN SIGHT INSTALLED.

Normal flight operation of the sight is accomplished as follows:

1. Gun safety switch at GUN.
2. Identify target; then set span adjustment lever to correspond with span of target airplane.
3. Fly airplane so that target appears within gyro reticle circle, and rotate throttle twist grip until diameter of gyro reticle circle corresponds to target size.
4. Frame target with gyro reticle circle by rotating twist grip as range changes. Track target smoothly for one second; then fire.

Note

The gyro sight computes correctly only after the target has been correctly framed and tracked for a minimum period of one second.

5. Continue ranging and tracking while firing.

N-9 GUN SIGHT.

Early airplanes are equipped with the N-9 gun sight (figure 4-13), which does not compute the lead for the pilot. The reticle image of this sight, although it is only a few inches from the pilot's eye, is made to look as if it were at a distance in front of the airplane. There is no need for constant readjustment of focus, and the sight may be viewed with both eyes open. The pilot may move his head in relation to the sight without disturbing the alignment of the target. A ball bank indicator is mounted on the N-9 sight, and the ball must be centered during gun firing. If the ball is to either side, the bullets will be off the target.

N-9 SIGHT CONTROLS.

SIGHT DIMMER RHEOSTAT. The rheostat for controlling the intensity of the reticle image is located below the center of the instrument panel, on the front switch panel, and has OFF and ON positions, with any desired position between.

SIGHT FILAMENT SWITCH. A sight filament switch (figure 4-13) is located on the left side of the sight and has an ALTER. FILAM'T position. If the light in the sight fails to go on, the filament selector switch should be moved to ALTER. FILAM'T.

FIRING GUNS WITH N-9 SIGHT INSTALLED.

Since the N-9 sight is a noncomputing reflex type with a fixed reticle, no control is provided other than the rheostat for reticle brilliance. The principle of operation is the apparent projection of the reticle image in space, which is similar to having the reticle image superimposed on the target. When the target is framed by the reticle, the guns may be fired by depressing the trigger on the control grip.

GUN CAMERA.

A Type N-4 or N-6 gun camera is installed in the leading edge of the left wing. The camera can be set for 16, 32, or 64 frames per second. The diaphragm stops are marked "B" (bright), "H" (hazy), and "D" (dull) to allow for various light conditions. On early airplanes, the camera opening is covered by glass. During take-offs and landings, the glass is likely to become scratched or obscured; therefore, on late airplanes, the camera opening is covered by a spring-loaded metal plate, which is actuated by a cable attached to the left landing gear. The plate opens when the gear is retracted and closes when the gear is lowered. A heater in the camera functions automatically when temperatures are low, provided the gun safety switch is at GUN or CAMERA AND SIGHT.

GUN CAMERA CONTROLS.

GUN SAFETY SWITCH. Refer to "Gunnery Equipment Controls" in this section.

TRIGGER. Refer to "Gunnery Equipment Controls" in this section.

Figure 4-13

Figure 4-14

BOMBING EQUIPMENT.

An external, removable bomb rack may be installed under each wing. Each rack will hold one 100-, 250-, or 500-pound bomb. Chemical tanks or combat fuel tanks may be carried on the bomb racks when bombs are not installed. The tanks are released either by normal or salvo operation of the bomb control system. Two bomb salvo handles (figure 4-14) permit selective mechanical release of bombs or tanks. The bomb system electrical controls consist of a bomb release switch on top of the control stick, three bomb arming switches, and a bomb release selector switch. Bombs are aimed before release by setting the gun sight selector switch at FIXED and using the gun sight.

BOMBING EQUIPMENT CONTROLS.

BOMB (TANK)-ROCKET SELECTOR SWITCH. The bomb (tank)-rocket selector switch (36, figure 1-6) is located below the left side of the instrument panel and has three positions: SAFE, BOTH, and TRAIN. With the selector switch at SAFE, bombs or tanks will not release when the bomb release button on the control stick is pressed. With the switch at BOTH, both bombs or drop tanks will release simultaneously when the bomb release button is pressed. With the switch at TRAIN, the left bomb or drop tank will drop when the bomb release button is pressed, and the right bomb or drop tank will drop when the button is pressed again. On some early airplanes, a SEL position replaces the TRAIN position. Airplanes with zero-rail rockets installed have the bomb-rocket selector switch located on the front switch panel. (See figure 4-9.) On these airplanes, the switch has ROCKETS, SAFE, BOTH, and TRAIN positions. The additional ROCKETS position permits the rockets to be fired when the rocket control panel switches are set as desired and the bomb-release button is pressed.

BOMB ARMING SWITCHES. The bomb arming switches (37, figure 1-6) are located below the left side of the instrument panel (figure 4-10) on late airplanes and on the front switch panel on airplanes with zero rail rockets installed. There are three arming switches: two nose-arming switches, marked "LEFT" and "RIGHT"; and a tail-arming switch. On early airplanes, all three of the switches have two positions, ON and OFF. When chemical tanks are carried on early airplanes, the left and right nose-arming switches must be moved to the ON position until smoke appears at the chemical tank and returned to OFF. On late airplanes, the two nose-arming switches have three positions: CHEM. RELEASE for use when chemical tanks are installed, NOSE ARM for use when bombs are installed, and OFF. The tail-arming switch has two positions, TAIL ARM and OFF. The OFF positions of the switches permit the bombs to be dropped safe.

CAUTION

The NOSE ARM position of the nose-arming switch on late airplanes must not be used when chemical tanks are installed, to prevent inadvertent discharge of chemicals.

BOMB-ROCKET RELEASE BUTTON. The bomb-rocket release button (figure 4-11) is located on the top of the control stick grip. Depressing the button fires the rockets or releases the bombs (or drop tanks) according to the positioning of the bomb-rocket selector switch.

RELEASING BOMBS.

The following procedure may be used to release bombs:

1. Move bomb arming switches to desired position for nose or tail arming.
2. Place bomb-rocket selector switch to TRAIN or BOTH position as desired.
3. Press bomb-rocket release button on control stick grip momentarily to release bombs. If bomb-rocket selector switch is at TRAIN, the bomb-rocket release button will release only the left bomb. Pressing the bomb-rocket release button again will release the bomb on the right rack.

Note

Bombs may be released when the airplane is in any pitch attitude from a 30-degree climb to a vertical dive.

Caution

Do not release the bombs when you are sideslipping more than 5 degrees in a vertical dive because of the danger of released bombs falling into the propeller.

4. After bomb release, move arming switches to OFF and bomb-rocket selector switch to SAFE.

EMERGENCY BOMB AND DROP TANK RELEASE.

Two bomb salvo handles (15, figure 1-4), located aft of the instrument panel on the left side of the cockpit, can be used to release the bombs or drop tanks manually in the event of failure of the normal electrical release. The two handles are mounted side by side and may be operated simultaneously with one hand. To drop bombs in safe condition, have bomb arming switches at OFF.

ROCKET EQUIPMENT.

Late airplanes are equipped to carry 10 zero-rail rockets attached to two pads on the underside of the wing. These rockets are aimed using the gun sight at FIXED position. If bombs or drop tanks are carried, only six rockets may be carried. The rocket armament switches are located on the front switch panel on a special rocket panel insert. (See figure 4-9.)

ROCKET EQUIPMENT CONTROLS.

BOMB-ROCKET SELECTOR SWITCH. For rocket firing, the bomb-rocket selector switch, located on the front switch panel, is used in conjunction with the rocket control panel (figure 4-9), and should be at ROCKETS. This completes the rocket-firing circuit. Setting the rocket control panel as desired releases rockets when the bomb-rocket release button on the control stick is pressed.

Note

When the bomb-rocket selector switch is at ROCKETS, the bomb release circuits are inoperative.

BOMB-ROCKET RELEASE BUTTON. For rocket firing, depressing the bomb-rocket release button (figure 4-11) on the control stick releases rockets either one at a time or all rockets in train, provided the bomb-rocket selector switch is at ROCKETS and the rocket control panel is set as desired.

ROCKET RELEASE CONTROL SWITCH. A rocket release control switch (figure 4-9), with OFF, SINGLE, and AUTO positions, is located on the front switch panel. When the switch is at AUTO, all the rockets fire in train when the bomb-rocket release button is depressed for approximately one second. Moving the switch to SINGLE allows one rocket to fire each time the bomb-rocket release button is depressed.

ROCKET DELAY SWITCH. A two-position rocket delay switch is located on the rocket panel. The switch is moved to DELAY to nose-arm the rockets for an instant's delay upon impact. The alternate switch position is INST.

ROCKET COUNTER CONTROL. A rocket counter is incorporated into the rocket panel and has a control knob located adjacent to its window. The counter shows the rocket to be fired by number and should be reset to 1 at the start of a mission.

Note

The firing order of the rockets singly or in train is as follows: 1, 3, 5, 7, and 9 on the left wing, 2, 4, 6, 8, and 10 on the right wing. (Rockets 7, 8, 9, and 10 are not installed when bombs are carried.)

FIRING ROCKETS.

To fire the rockets, the following sequence should be followed:

1. Turn rocket counter dial to 1.
2. Place bomb-rocket selector switch in ROCKETS position.

Note

When this switch is in the ROCKETS position, the bomb release circuits are inoperative.

3. To nose-arm rockets for delay upon impact, turn rocket delay switch to DELAY.
4. To fire rockets singly, move rocket release control switch to SINGLE and press bomb-rocket release button on control stick once for each rocket.
5. To fire all rockets in train, move rocket release control switch to AUTO and hold bomb-rocket release button depressed for approximately one second.

CHEMICAL TANK EQUIPMENT.

Chemical tanks may be installed in lieu of bombs on the bomb rack on each wing. To release chemicals from the chemical tanks on early airplanes, place nose-arming switches at ON position until smoke appears; then return switches to OFF. On late airplanes, chemicals are released by holding the nose-arming switches up at CHEM. RELEASE until smoke appears, then releasing them to OFF. These switches are spring-loaded to return to OFF from the CHEM. RELEASE position. The chemical tanks are released in the same manner as bombs during normal and emergency release.

MISCELLANEOUS EQUIPMENT.

ANTI-G SUIT PROVISION.

An air pressure outlet connection on the left side of the pilot's seat provides for attachment of the air pressure intake tube of the anti-G suit. Air pressure for the inflation of the anti-G suit bladders is supplied from the exhaust side of the engine-driven vacuum pump and is regulated by a Type M-2 valve, which also regulates pressure to the drop tanks. If drop tanks are installed on the airplane, the acceleration force (G-load) required to actuate the M-2 valve should be approximately 3 to 3½ G because of the approximately 5 psi pressure exerted in the tanks. If drop tanks are not installed, the valve should open at 2 G. After the valve opens, pressure is passed through a regulator valve into the suit in proportion to the G-force imposed. For every 1 G acceleration force, a corresponding one psi air pressure is exerted in the anti-G suit.

DATA CASE.

A data case is fastened to the access door on the underside of the fuselage, just forward of the tail wheel. (See figure 1-3.)

TAIL POSITION LIGHT LENSES.

On the late airplanes, a case containing three tail position light lenses (red, green, and clear) is accessible through the data case door.

MAP CASE.

A map case (26, figure 1-4) is located to the left of the seat.

DROP MESSAGE BAG.

A drop message bag is contained in a holder (27, figure 1-4) on the map case cover.

ENGINE CRANK.

Early airplanes have an engine crank and extension tube stowed in brackets at the back of the right main landing gear well. On late airplanes, these parts are not included.

ARMREST.

A folding armrest is located on the left longeron, aft of the engine control quadrant.

RELIEF TUBE.

A relief tube is stowed on a bracket on the floor of the cockpit, at the left of the pilot's seat.

AIRPLANE TIE-DOWN.

Tie-down points are provided on each wing, each main wheel axle, and the fuselage. A flush mooring ring is provided on the lower surface of each wing, approximately in line with the outboard end of the wing flap. These rings are pried out for use. A mooring ring is provided on the inboard side of each main landing gear axle. For fuselage tie-down, the tie-down rope is passed through the lift tube below and aft of the insignia.

Operating Limitations

SECTION V

INSTRUMENT MARKINGS.

Instrument markings showing the various operating limits are illustrated in figure 5-1. In some cases, the markings represent limitations that are self-explanatory and therefore are not discussed in the text. Operating restrictions or limitations that do not appear as maximum limits on the cockpit instruments are discussed in detail in the following paragraphs. Limitations relative to hot- and cold-weather operation, instrument flight, and flying through turbulent air are covered in Section IX.

ENGINE LIMITATIONS.

All normal engine limitations are shown in figure 5-1. The maximum diving engine overspeed is 3300 rpm.

WARNING

Whenever engine speed exceeds operating limits, airplane should be landed immediately at the nearest base. The reason for the overspeed (if known), the maximum rpm, and duration must be entered in the Form 1 and reported to the maintenance officer. Overspeed between 3300 and 3600 rpm necessitates an inspection of the engine before further flight. If the rpm exceeds 3600 rpm, the engine must be removed for overhaul.

Avoid operation below 1600 rpm in low blower. Avoid operation below 2000 rpm in high blower.

AIRSPEED LIMITATIONS.

The red pointer on the airspeed indicator marks the maximum permissible airspeed (505 mph) up to 9000 feet. For maximum airspeed permissible above 9000 feet, see figure 6-3. Do not exceed the following wing flap setting airspeed restrictions:

ANGLE DOWN (DEGREES)	MAXIMUM IAS (MPH)
10	400
20	275
30	225
40	180
50	165

When droppable 75-gallon combat fuel tanks are installed, do not exceed 400 mph indicated airspeed. Do not allow airspeed to fall below 110 mph indicated airspeed during a sideslip.

INSTRUMENT *Markings*

TACHOMETER

(white)	1600-2700 rpm	Continuous
(white)	2400 rpm	Maximum Cruise
(white)	2700 rpm	Maximum Continuous
(red)	3000 rpm	Take-off, Military, and War Emergency Power
(red)	3300 rpm	Maximum Diving Overspeed

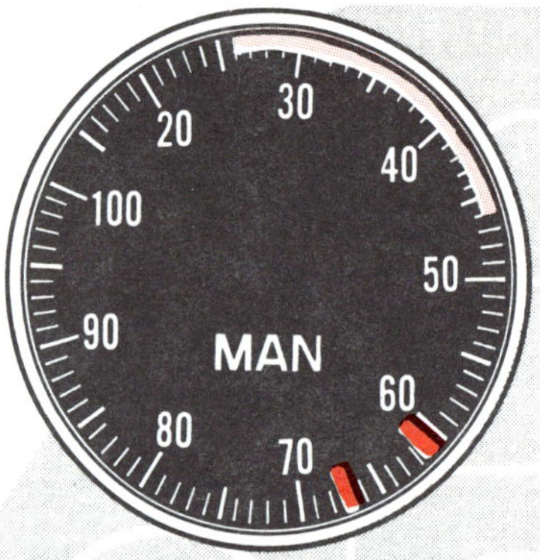

MANIFOLD PRESSURE

(white)	26-46 in. Hg	Desired Operating Range
(red)	61 in. Hg	Take-off and Military Power
(red)	67 in. Hg	War Emergency (5 Min)

CARBURETOR AIR TEMP

(white)	15°C to 40°C	Normal
(pink)	-10°C to +15°C	Caution (Danger of Icing)
(red)	50°C	Maximum (Danger of Detonation)

COOLANT TEMPERATURE

(red)	60°C	Minimum Take-off
(white)	100°C-110°C	Normal
(red)	121°C	Maximum
(red)	125°C	Maximum (-9 Engines)

109-51-648

Figure 5-1 (Sheet 1 of 2)

OIL TEMPERATURE

20°C	Minimum
70°C-80°C	Normal
105°C	Maximum

OIL PRESSURE

50 psi	Minimum
70-80 psi	Normal

FUEL PRESSURE

16 psi	Minimum
16-18 psi	Normal
20 psi	Maximum

ENGINE GAGE UNIT

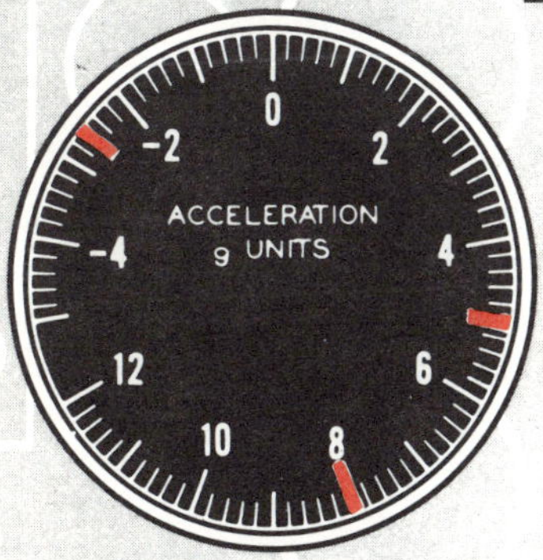

ACCELEROMETER

5 G	Max at Gross Weight of 10,840 Lb (Two 75 Gal Drop Tanks)
8 G	Max at Design Gross Weight (8000 Lb)
−2.6 G	Max at Gross Weight of 10,840 Lb (Two 75 Gal Drop Tanks)

AIRSPEED

Max Permissible IAS With Flaps Full Down — 165 mph
(Max Permissible IAS With Gear Extended — 170 mph)

The instrument setting is such that the red pointer will move to indicate the limiting structural airspeed of 505 mph or the airspeed representing the limiting Mach No. of .77, whichever is less.

HYDRAULIC PRESSURE

800-1250 psi	Normal (System Pressurized)
600 psi	Caution (Min for Operation)
1250 psi	Maximum
0-150 psi	Normal (System Depressurized)

SUCTION

3.75 in. Hg	Minimum
3.75-4.25 in. Hg	Normal
4.25 in. Hg	Maximum

109-51-657

Figure 5-1 (Sheet 2 of 2)

COMPRESSIBILITY EFFECTS.

At high diving speeds, there is danger of the airplane being affected by compressibility as the airplane speed approaches the speed of sound. Compressibility effects are indicated by instability of the airplane, uncontrollable rolling or pitching, stiffness of controls, or combinations of these effects. A nose-heaviness will be noticed and will become more severe as speed increases.

PROHIBITED MANEUVERS.

No intentional power-on spins or snap rolls are permitted, as it is impossible to do a good snap roll and most attempts end up in a power spin.

No intentional power-off spins are permitted below 12,000 feet.

Abrupt pull-ups should not be attempted with more than 25 gallons in the fuselage fuel tank.

No aerobatics are permitted with fuel in the fuselage tank.

No aerobatics are permitted when drop tanks are installed.

No aerobatics are permitted if 1000-pound bombs are installed.

Inverted flying must be limited to 10 seconds because of loss of oil pressure and failure of the scavenge pumps to operate in an inverted position.

ACCELERATION LIMITATIONS.

The airplane is limited to a maximum positive load factor of 8 G and a maximum negative load factor of 4 G. These limits apply only when the clean airplane gross weight does not exceed 8000 pounds (design gross weight). When airplane gross weight is greater than 8000 pounds, the maximum allowable G is less than the maximum limit marked on the accelerometer. Remember that when you pull the maximum G, the wings of your airplane must support eight times their normal load. This means that during a maximum G pull-out, the wings of the airplane (at design gross weight) are supporting eight times 8000 pounds or a total of approximately 64,000 pounds (the maximum that the wings can safely support). Therefore, when your airplane weighs more than 8000 pounds, the maximum G that you can safely apply can be determined by dividing 64,000 by the new gross weight. When external loads are carried, the maximum allowable G-load is 5 G. The maximum load factors we have been talking about apply only to straight pull-outs. Rolling pull-outs are a different story, however, since they impose considerably more stress upon the airplane. The maximum allowable load factor in a rolling pull-out is two-thirds the maximum G for a straight pull-out.

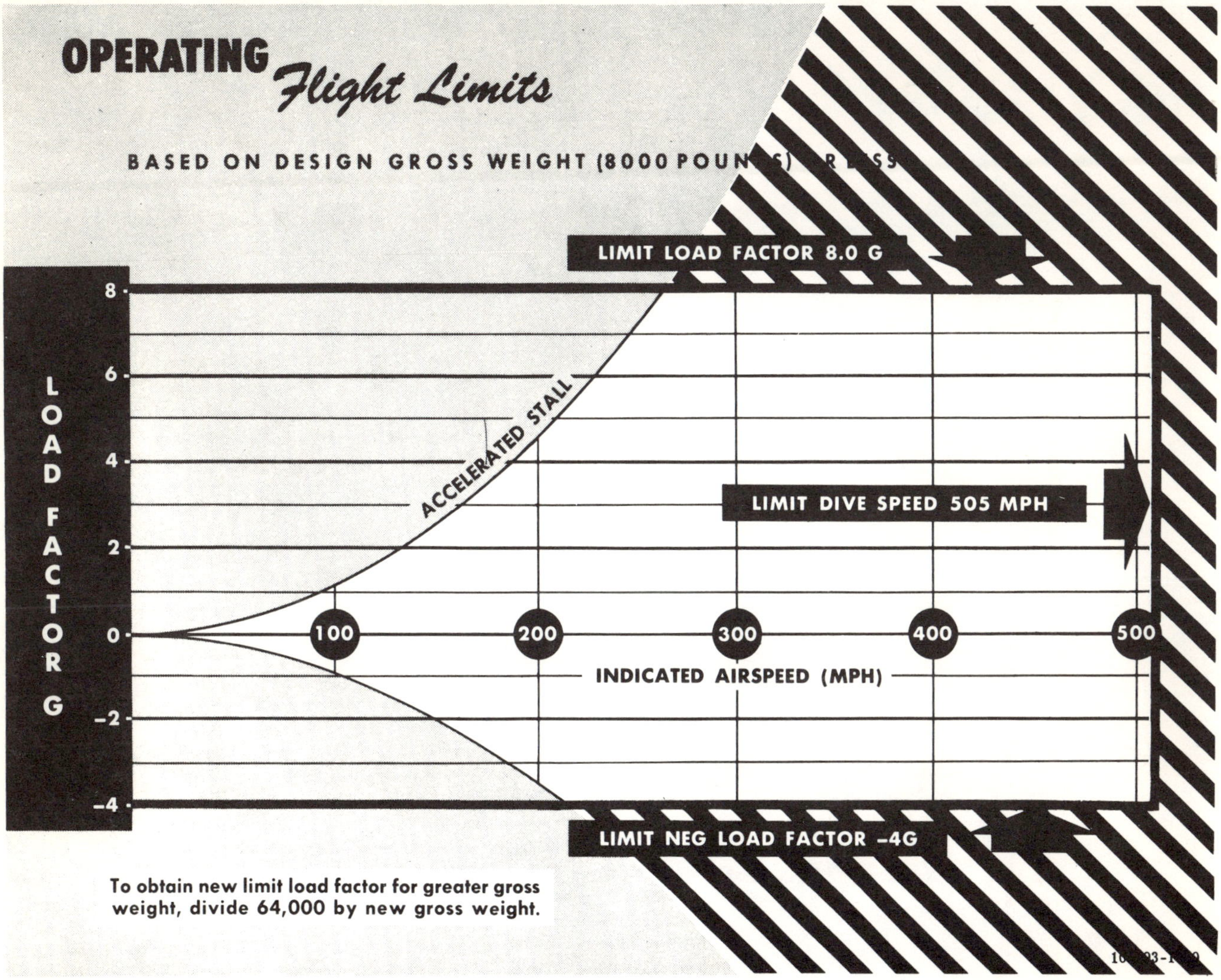

Figure 5-2

OPERATING FLIGHT LIMITS.

The Operating Flight Limits chart (figure 5-2) shows the G-limitations of the airplane. Various load factors are shown vertically along the left side of the chart, and various indicated airspeeds are shown horizontally across the center of the chart. The horizontal red lines at the top and bottom of the chart represent the maximum positive and maximum negative allowable load factors. The vertical red line indicates the limit dive speed of the airplane. The curved lines show the G at which the airplane stalls at various airspeeds. The upper curved line shows, for example, that at 150 mph the airplane stalls in a 2.5 G turn, while at 200 mph the airplane does not stall until more than 4.5 G is applied. The upper and lower limits at the right side of the chart show that the maximum positive and negative limit load factors (+8 G and –4 G) can be safely applied up to the limit dive speed of the airplane.

CENTER-OF-GRAVITY LIMITATIONS.

Any configuration of external load that the airplane is designed to carry may be installed without exceeding the CG limits. To prevent the possibility of an adverse aft CG condition, which would cause unsatisfactory flight characteristics, the fuselage fuel tank capacity is restricted to 65 gallons maximum. No flight should be permitted with more than 65 gallons in the fuselage fuel tank because of the possibility of a stick force reversal during an accelerated maneuver such as a dive pull-out. A forward CG condition exists when less than 25 gallons of fuel remains in the fuselage fuel tank. In this case, landings should be performed with caution, particularly during flare-out and immediately after touchdown, to prevent nosing over.

WEIGHT LIMITATIONS.

There are no weight limitations to observe, since the external mounting provisions prevent overloading.

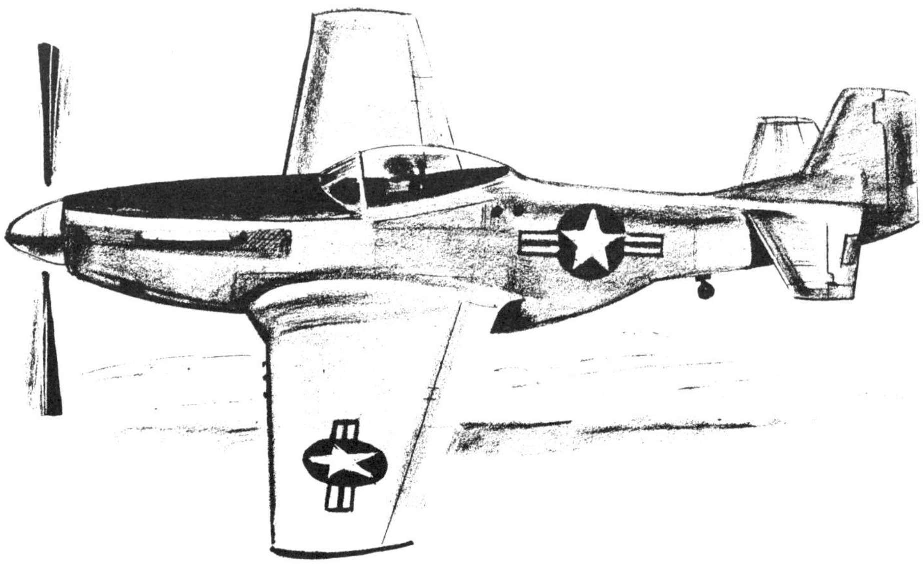

Flight Characteristics

SECTION VI

GENERAL.

The airplane is stable at all normal loadings. The directional trim changes as speed and horsepower output are varied. The trim tab controls are sensitive and must be used very carefully to prevent overtrimming. An elevator bobweight in the elevator control system artificially loads the controls to help prevent overaccelerating during tight turns or pull-outs because of the light stick forces.

STALLS.

The airplane has a comparatively mild stall. The airplane doesn't whip at the stall, but rolls rather slowly and has very little tendency to drop into a spin. When you release the stick and rudder, the nose drops sharply and the airplane recovers from the stall almost instantly. When a complete stall is reached, a wing drops. If you keep pulling back on the stick when the wing drops, the airplane falls into a steep spiral. In a straight power-off stall, some warning is given about 3 to 4 mph above the stall by a slight elevator buffet. A high-speed stall is preceded by a sharp buffeting at the elevators and wing root, but recovery is almost immediate when pressure on the stick is released. The normal procedure holds good when recovering from any stall. Release the back pressure on the stick and pick up the dropping wing with opposite rudder. The speed at which a stall occurs can vary widely, depending on the gross weight. (See figure 6-1.)

Note

> The airplane effects its own recovery from a stall in a slow-speed turn by performing a partial snap roll and stopping in level position. From a stalled turn, the airplane always rolls to the outside of the turn.

PRACTICE STALLS.

The following practice stalls will acquaint you with the stall traits and stall speed of the airplane under various flight conditions. For both power-on and power-off stalls, set propeller control to obtain 2700 rpm. Retard throttle smoothly to 10-12 in. Hg for power-off stalls; set manifold pressure at 30 in. Hg for power-on stalls.

PRACTICE STALL—GEAR AND FLAPS DOWN, POWER OFF, STRAIGHT AHEAD.

1. Close throttle.
2. Gear down at 170 mph.

Stall Speeds

IAS • MPH (POWER OFF)

BASED ON FLIGHT TESTS

	GROSS WEIGHT LB	GEAR UP FLAPS UP			GEAR DOWN FLAPS 45° DOWN		
		LEVEL	30° BANK	45° BANK	LEVEL	30° BANK	45° BANK
WITH WING RACKS	10,000	106	115	128	101	110	123
	9,000	101	109	121	94	103	116
	8,000	94	102	114	87	98	108
WITH BOMBS, DROP TANKS, OR ROCKETS*	12,000	119	128	143	113	123	136
	11,000	113	122	137	107	117	131
	10,000	108	116	130	102	111	124
	9,000	102	110	123	95	105	117

*STALL SPEEDS WITH ROCKETS ARE ESTIMATED.

109-93-1730

Figure 6-1

3. Lower full flaps at 160 mph.

4. Establish 130 mph glide and raise nose to landing attitude.

5. Hold this attitude until stall breaks; observe characteristics of airplane in stall (usually the left wing stalls before the right wing). After nose drops, initiate stall recovery by smoothly advancing throttle to 45 in. Hg and easing stick forward to regain flying speed. Level wings with rudder and aileron and regain 130 mph; then reduce throttle to 30 in. Hg.

6. Raise gear and flaps.

PRACTICE STALL—GEAR AND FLAPS DOWN, POWER OFF, MEDIUM BANK.

1. Close throttle.
2. Gear down at 170 mph.
3. Lower full flaps at 160 mph.
4. Establish 130 mph glide.
5. Establish a medium bank right or left; slow airplane and tighten turn with elevator until stall breaks.
6. As stall breaks, recover with stick forward, and advance throttle smoothly to 45 in. Hg.
7. Roll wings level with rudder and aileron as soon as possible.
8. Raise gear and flaps; return to cruising power.

PRACTICE STALL—GEAR AND FLAPS UP, POWER ON, STRAIGHT AHEAD OR IN TURN.

1. Cruise throttle setting.
2. Raise nose to about a 40-degree climb attitude straight ahead, or use a gentle climbing turn right or left and tighten turn with back pressure until stall breaks.
3. As stall breaks, effect normal recovery, advancing throttle smoothly to 45 in. Hg.
4. Retard throttle to cruise power after recovery.

SPINS.

POWER-OFF SPINS.

For spin characteristics, see figure 6-2. In general, spins in this airplane are uncomfortable because of heavy oscillations. Occasionally, the left spin oscillations will dampen out after approximately three turns, but the right spin oscillations will not. When controls are applied to start a spin, the airplane snaps one-half turn in the direction of spin, with the nose dropping to near vertical. At the end of one turn, the nose rises to or above the horizon and the spin slows down, occasionally coming almost to a complete stop. The airplane then snaps one-half turn with the nose dropping to 50-60 degrees below the horizon and continues as during the first turn. The force required to hold the controls in the spinning position is quite heavy, and some rudder buffet becomes noticeable. When controls are applied for recovery, the nose drops to near the vertical position and the spin speeds up and then stops in one to 1¼ turns.

POWER-OFF SPIN RECOVERY.

Recovery procedure is the same in both a left and a right spin. As soon as you apply opposite rudder, the nose drops slightly. The spin speeds up for about 1¼ turns and then stops. The rudder force is light at first, becomes very heavy for about one second in the first one-half turn, and then drops to zero as the spin stops. Recovery is effected in the normal manner as follows:

1. Controls with spin.
2. Apply full opposite rudder.
3. Stick neutral after airplane responds to rudder (as rotation stops).
4. Rudder to neutral and complete recovery as spin ends.

Note

> During the spin, a slight rudder buffeting is noticeable. If you attempt to recover from the dive too soon after the spin stops, you will also feel a rather heavy buffeting in both the elevator and rudder. The remedy for this condition is to release some of the back pressure on the stick.

POWER-ON SPINS.

Power-on spins should never be intentionally performed in this airplane. In a power-on spin, the nose of the airplane remains 10 to 20 degrees above the horizon, and recovery control has no effect upon the airplane until the throttle is completely retarded.

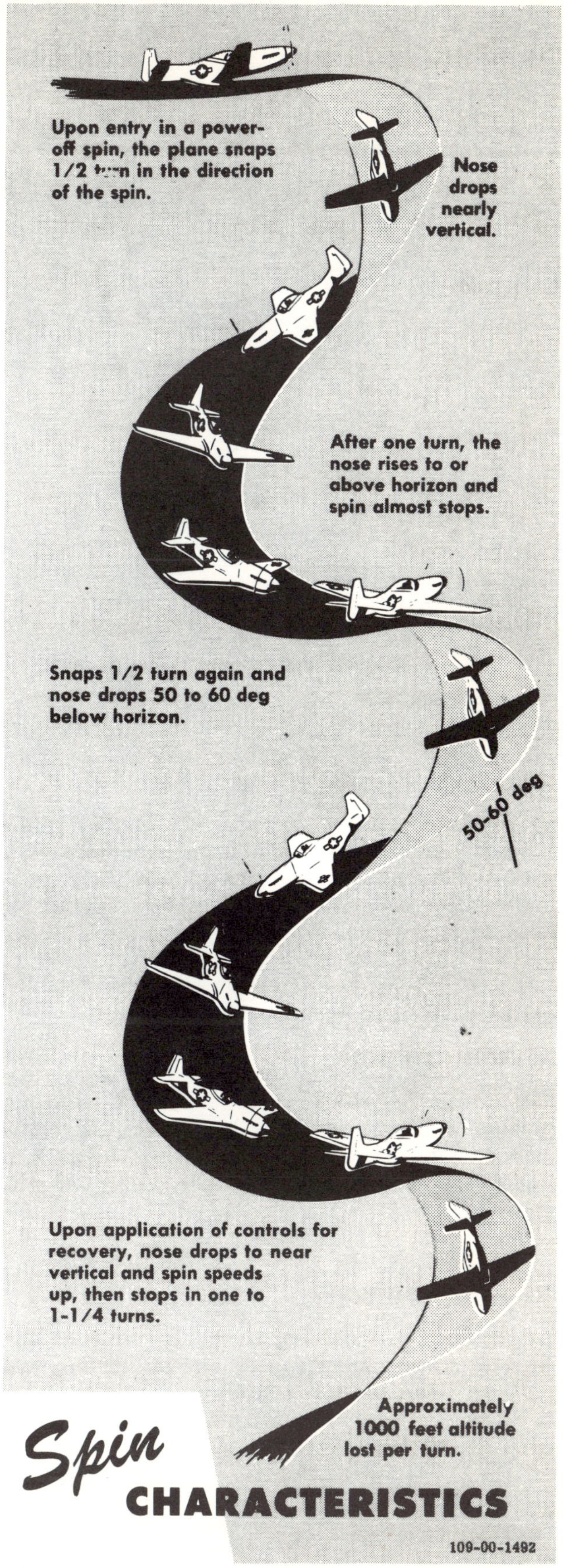

Figure 6-2

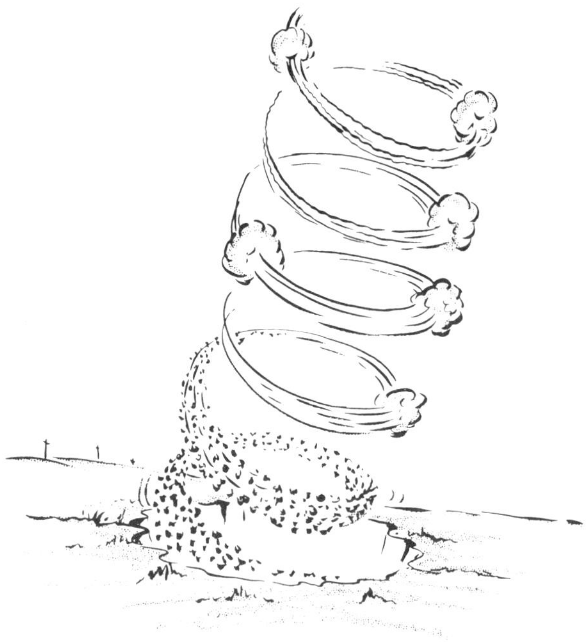

Warning

Power-on spins are extremely dangerous in this airplane.

POWER-ON SPIN RECOVERY.

Close throttle completely and apply controls as for power-off spin recovery. Hold full opposite rudder with stick in neutral until recovery is effected. As many as five or six turns are made after rudder is applied for recovery, and 9000 to 10,000 feet of altitude is lost.

FLIGHT CONTROL EFFECTIVENESS.

AILERON CONTROL.

The airplane has sealed-balance ailerons. A fabric diaphragm seals the space between the leading edge of the aileron and the aft side of the wing spar, and tends to lighten the stick forces. At normal speeds, control is positive.

ELEVATOR CONTROL.

At normal speeds, elevator control is very good and stick pressure is light. As speed increases in a dive and pull-out is attempted, you must use caution so as not to overcontrol and pull up too fast; otherwise, undue stress or even failure may result.

RUDDER CONTROL.

Because of the reverse-boost rudder tab and dorsal fin, the airplane has very good directional stability, with a directional change requiring definite pressure on the rudder pedal in proportion to the amount of yaw desired.

TRIM TAB CONTROL.

Trim tabs are very sensitive and should be used with care.

LEVEL-FLIGHT CHARACTERISTICS.

LEVEL-FLIGHT STABILITY.

Level-flight stability is good, with normal control pressure required for desired effect.

LOW SPEED.

At low speed, the controls tend to become slightly mushy but control is still good.

CRUISE SPEED.

Controls are positive at this speed, and control stick forces are normal.

MANEUVERING FLIGHT.

MANEUVERING-FLIGHT STABILITY.

Combat maneuvers should never be attempted when the fuselage tank contains more than 25 gallons of fuel, as the tail-heavy condition could cause a reversal of control stick forces during any abrupt maneuver approaching 6.75 G. At this point, forward pressure is needed to prevent the pull-out or turn from tightening up to the point where structural failure results. With the fuselage tank empty, stick forces are normal with slightly less pull force required above 6 G. However, a positive back pressure is needed up to maximum G-load.

MANEUVERABILITY.

Control response for this airplane is very good, with unusually light stick pressures required to perform maneuvers. The reverse-boost rudder trim tab gives the desired effect, that an increase in rudder pedal pressure is always necessary to obtain an increase in yaw angle.

CAUTION

Avoid rapid movement of the elevators in dives and maneuvers at speeds above 350 mph IAS or .7 Mach, particularly at aft CG positions, to prevent rapid uncontrolled increase in G-loads.

DIVES.

DIVES AND ACCELERATED FLIGHT.

At high diving speeds, there is danger of the airplane being affected by compressibility—a phenomenon likely to be encountered when speed approaches the speed of sound. Compressibility may be indicated by instability of the airplane, uncontrollable rolling or pitching, stiffness of controls, or a combination of these effects. The longitudinal characteristics remain normal until the speed of the airplane reaches approximately 72 to 74 percent of the speed of sound. At this speed, the airplane may become slightly nose-heavy because of the compressibility. Inasmuch as further increases in speed may result in more severe nose-heaviness, diving speed should be limited at this point and recovery started immediately after the change in longitudinal trim is evident.

ALTITUDE REQUIRED FOR PULL-OUT.

Figure 6-3 shows the altitude used in a pull-out from dives with a constant 4 G or 6 G acceleration.

WARNING

The anti-G suit should be used with a constant 6 G pull-out.

DIVE RECOVERY.

If diving speed limits are exceeded, compressibility effects will be experienced. Reduce power and pull up very gradually, taking care not to exceed 4 G except in emergencies. If pull-out above 4 G is necessary, as G increases, relax pull force on stick. The elevator trim tab is not normally required to aid recovery. In the event it is necessary, use with extreme care and feed in gradually.

WARNING

Care should be taken in pull-outs above 4 G, as the stick forces are relatively light, and an abrupt pull-out may cause rapid uncontrolled increase in G.

FLIGHT WITH EXTERNAL LOADS.

DROP TANKS.

At high speed (in excess of 400 mph), buffeting will be encountered if the airplane is carrying 75-gallon drop tanks.

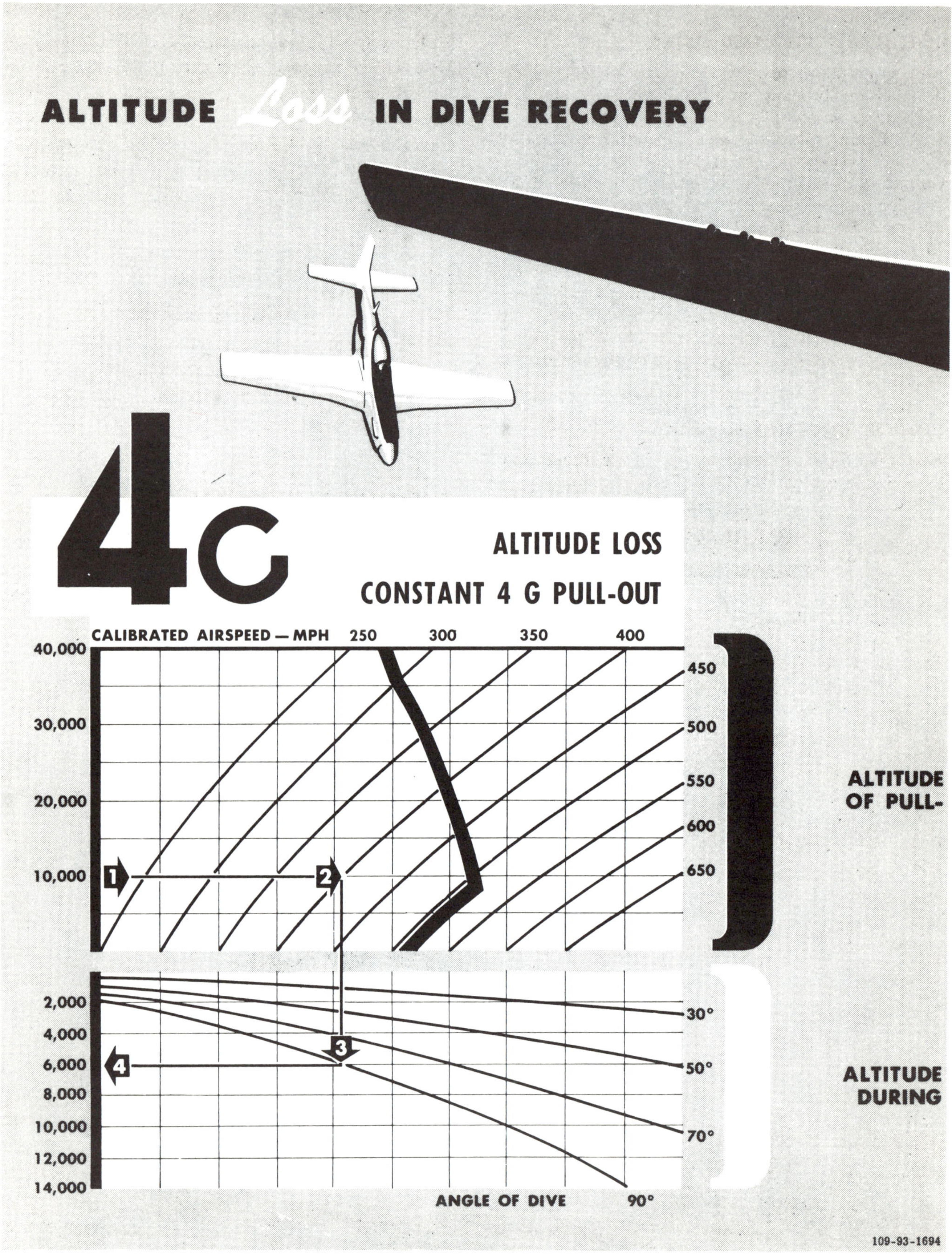

Figure 6-3 (Sheet 1 of 2)

HOW TO USE CHARTS

Select appropriate chart, depending upon acceleration (4 G or 6 G) to be held in pull-out; then —

1. Enter chart at altitude line nearest actual altitude at start of pull-out (for example, 10,000 feet).
2. On scale along altitude line, select point nearest the CAS at which pull-out is started (400 mph).
3. Sight vertically down to point on curye of dive angle (90°) directly below airspeed.
4. Sight back horizontally to scale at left to read altitude lost during pull-out (4 G — 6000 feet; 6 G — 4000 feet).

6G

ALTITUDE LOSS CONSTANT 6 G PULL-OUT

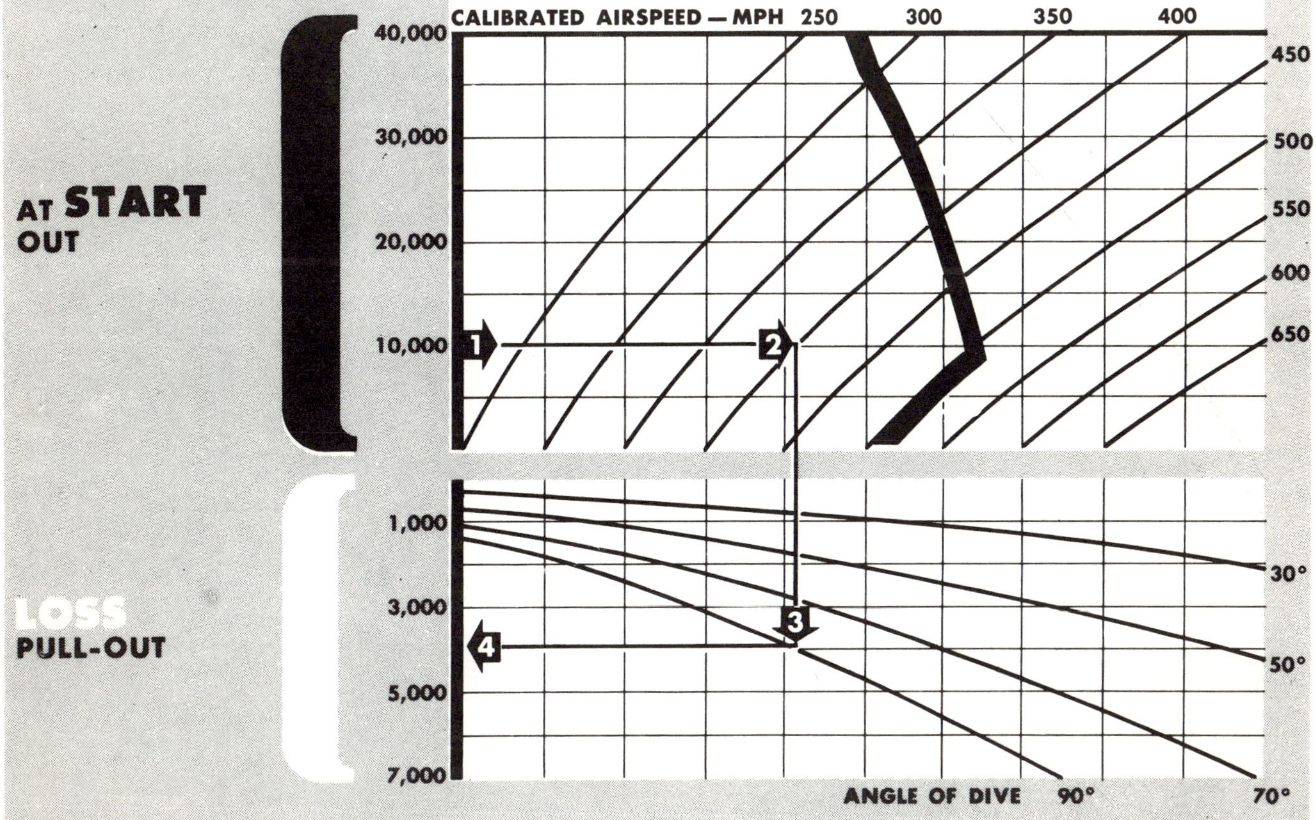

Figure 6-3 (Sheet 2 of 2)

Systems Operation

SECTION VII

ENGINE.

USE OF TAKE-OFF (MILITARY) POWER.

It is often asked what the consequences will be if the 5-minute limit at Take-off Power is exceeded. Another frequent inquiry is how long a period must be allowed after the specified time limit has elapsed until Take-off Power can again be used. These questions are difficult to answer, since the time limit specified does not mean that engine damage will occur if the limit is exceeded. Instead, the limit means that the total operating time at high power should be kept to a reasonable minimum in the interest of prolonging engine life.

It is generally accepted that high-power operation of an engine results in increased wear and necessitates more frequent overhaul than low-power operation. However, it is apparent that a certain percentage of operating time must be at full power. The engine manufacturer allows for this in qualification tests in which much of the running is done at Take-off Power to prove ability to withstand the resulting loads. It is established in these runs that the engine will handle sustained high power without damage. Nevertheless, it is still the aim of the manufacturer and to the best interest of the pilot to keep within reasonable values the amount of high-power time accumulated in the field. The most satisfactory method for accomplishing this is to establish time limits that will keep pilots constantly aware of the desire to hold high-power periods to the shortest period that the flight plan will allow, so that the total accumulated time and resulting wear can be kept to a minimum. How the time at high power is accumulated is of secondary importance; i.e., it is no worse from the standpoint of engine wear to operate at Take-off Power for one hour straight than it is to operate in twelve 5-minute stretches, provided engine temperatures and pressures are within limits. In fact, the former procedure may even be preferable, as it eliminates temperature cycles which also promote engine wear. Thus, if flight conditions occasionally require exceeding time limits, this should not cause concern so long as constant effort is made to *keep the over-all time at Take-off Power to the minimum practicable.*

Another factor to be remembered in operating engines at high power is that full Take-off Power (3000 rpm and 61 in. Hg) is to be preferred over take-off rpm with reduced manifold pressure. This procedure results in less engine wear for two reasons. First, the higher resulting brake horsepower decreases the time required to obtain the objective of such high-power operation. At take-off, for example, the use of full power decreases the time required to reach an altitude and airspeed where it is safe to reduce power and shortens the time required to reach the airspeed that will provide more favorable cooling. Second, high rpm results in high loads on the reciprocating parts because of inertia forces. As these loads are partially offset by the gas pressure in the cylinder, the higher cylinder pressures resulting from use of full take-off manifold pressure give lower net loads and less wear. Sustained high rpm is a major cause of engine wear. It requires more "rpm minutes" and "piston-ring miles" to take off with reduced manifold pressure. In addition to the engine wear factor, taking off at reduced power is comparable to starting witb approximately one-third of the runway behind the airplane. Therefore, full power should *always* be used on take-offs.

WAR EMERGENCY POWER.

During emergencies in a combat zone, it is sometimes necessary to get the absolute maximum manifold pressure at which the engine may be operated within reasonable safety limits. This extra power is available when the throttle is pushed beyond a gate on the throttle quadrant, provided the following requirements are fulfilled.

1. The airplane must be placarded with a decal stating that use of War Emergency Power is permitted.

2. Fuel Grade 100/130 must be used, and a special type of spark plugs must be installed.

CAUTION

If the oil has been diluted, it is desirable to operate the engine 10 to 15 minutes at from 80 percent Normal Power to Military Power before using War Emergency Power, to remove excess fuel from oil.

SPARK PLUG FOULING.

Engine roughness may be the first sign of spark plug lead fouling, but to determine whether the plugs are at fault, clean out engine by advancing propeller control to 3000 rpm and throttle to 61 in. Hg and run engine continuously for one minute. Return throttle and propeller control to cruise setting and notice whether roughness persists. If roughness is still present, check for carburetor ice; then, if engine is still rough, reduce power to best operating setting and proceed to nearest base for landing to determine trouble.

Note

Entry must be made on Form 1 of the length of time of War Emergency Power operation, which is limited to a maximum period of 5 minutes.

Note

During prolonged cruising flight, "clean out" engine every 30 minutes. Also "clean out" engine before landing.

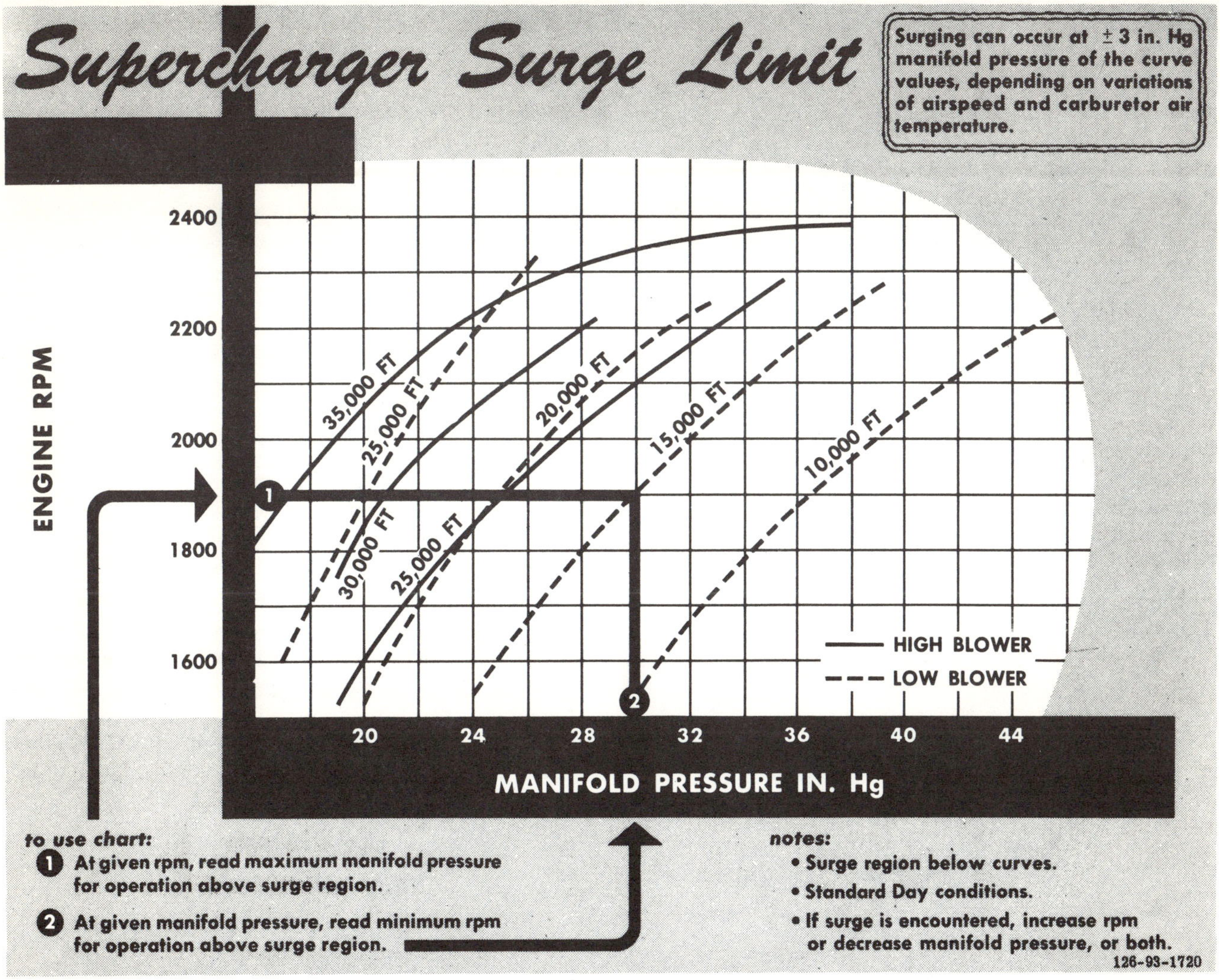

Figure 7-1

SUPERCHARGER SURGE.

Because of the design of the supercharger, surging may be encountered in high or low blower at various rpm, manifold pressure, and altitude combinations. Supercharger surging results when the airflow through the supercharger stalls; this causes a fluctuation in manifold pressure and induces erratic fuel metering. Under severe surging conditions, the engine cuts out completely, When surging is encountered, it may be corrected by either increasing the rpm setting or decreasing the manifold pressure setting. Typical supercharger surge curves for the V-1650-3 and V-1650-9A engines are shown in figure 7-1; this data is not available for the V-1650-7 engine, but the general shape of the curves is similar. The curves show the various manifold pressure and rpm combinations as which surging occurs at various altitudes.

CARBURETOR ICING.

Carburetor ice forms more readily when carburetor air temperature is between −10°C and +15°C. However, carburetor ice can form at any time, even with outside temperature as high as 32°C (90°F) and with temperature and dew point spread as much as 12°C (22°F). The formation of carburetor ice is hard to detect, since the automatic manifold pressure regulator maintains a constant manifold pressure. The only warning of carburetor ice is a roughness in the engine. If application of carburetor hot air does not remove roughness, clean out engine as directed under "Spark Plug Fouling" in this section. If carburetor ice is the cause of roughness, use hot air as needed to prevent further carburetor ice. If the air duct becomes obstructed with ice, hot air is automatically admitted to the air duct regardless of the position of the carburetor air control.

DETONATION.

Detonation is the result of one type of abnormal combustion of the fuel-air mixture. The other prevalent form of abnormal combustion is preignition. When detonation occurs, combustion is normal until approximately 80 percent of the charge is burning. At that point, the rate of combustion speeds up tremendously, resulting in an explosion or nearly instantaneous combustion. This explosion actually pounds the cylinder walls, producing knock. This knock, or pounding of the cylinder walls, can cause an engine failure. In an airplane, the knock is not heard because of other engine and propeller noises. However, detonation can be detected by observing the exhaust for visible puffs of black smoke, glowing carbon particles, or a small, sharp, whitish-orange flame. In addition, a rapid increase in coolant temperatures often indicates detonation. When detonation is evident, throttle reduction is the most immediate and surest remedy. *When detonation occurs, power is lost.* Contributing causes of detonation are as follows:

1. Low-octane fuel.
2. High coolant temperature caused by too long a climb at too low an airspeed or by too lean a mixture.
3. High mixture temperature caused by use of carburetor heat or by high outside air temperature.
4. Too high manifold pressure with other conditions favorable to detonation.
5. Improper mixture caused by faulty carburetor or too lean a mixture.

PREIGNITION.

Preignition is closely related to detonation. In fact, detonation often progresses into preignition. When the engine gets too hot, the mixture is ignited before the spark occurs. When this happens, much of the power is wasted trying to push the piston down while it is still rising in the cylinder. The power impulses are uneven, horsepower falls off, and the engine can be damaged from excessive pressures and temperatures. Preignition may be indicated by backfiring through the carburetor and possibly by a rapid increase in coolant temperature. When preignition is encountered, the throttle setting should be reduced immediately, as in severe cases, complete piston, valve and/or cylinder destruction can occur in a matter of a few seconds.

FUEL SYSTEM.

FUEL TANK SEQUENCE.

Take-off and climb should be accomplished with fuel tank selector handle at MAIN TANK L.H. because the vapor separation line from the carburetor returns to this tank. Fuel vapors flow from the carburetor to this tank at the rate of approximately one gallon per hour. At altitude, use fuel from fuselage fuel tank until 25 gallons remains, to have ideal CG condition for landing. Then cruise on drop tanks alternately until they are empty. Continue flight using both main wing tanks alternately, to prevent wing heaviness, until they are empty. Then use fuel from fuselage tank during landing.

WARNING

The fuel booster pump switch must be ON during flight to ensure an adequate fuel supply.

Note

For transitional training flights, keep approximately 25 gallons of fuel in fuselage tank to keep CG of airplane in optimum position for landing.

All-Weather Operation

Except for some repetition necessary for emphasis or continuity of thought, this section contains only those procedures that differ from, or are in addition to, the normal operating instructions contained in Section II.

INSTRUMENT FLIGHT PROCEDURES

Flying the airplane in all weather conditions requires proper instrument proficiency on the part of the pilot and thorough preflight planning. All the necessary flight instruments are provided, including directional gyro and flight indicator. Because of higher power settings and gradual letdowns, range is somewhat reduced.

Note

All turns are single-needle-width standard rate (3 degrees per second).

BEFORE TAKE-OFF.

Complete all checks required for any normal flight, with the following additions:

1. Check to be sure you have LF-MS edition (Radio Facilities Charts), AN 08-15-2 (USAF Radio Data and Flight Information), and T.O. No. 1F-51D-1 (Flight Handbook, formerly AN 01-60JE-1).
2. Check clock and set it to correct time.
3. Check suction gage for proper indication.
4. Check that pitot head cover has been removed. Turn pitot heater switch ON and have outside observer verify its operation. Then turn pitot heater switch OFF until airplane is in air, as there is insufficient cooling for pitot head while airplane is on ground.
5. Check airspeed needle at zero. Check airspeed correction card for any deviation at speed range to be flown.
6. The directional gyro rotor requires at least 5 minutes to attain proper operating speed. The dial card should revolve with the knob when the gyro is caged, but not when the gyro is uncaged. Set directional gyro so that it corresponds to reading of remote compass.
7. The gyro horizon rotor requires at least 5 minutes to attain proper operating speed. Cage instrument and uncage it. After the instrument is uncaged, the horizon bar should return to the correct position for the airplane in a three-point attitude. Temporary vibration of the horizon bar is permissible.

Note

If the horizon bar temporarily leaves the horizontal position while the airplane is being taxied straight ahead, or if the bar tips more than 5 degrees during taxiing turns, the instrument is not operating properly.

8. Obtain station altimeter setting (sea-level barometric pressure) from control tower operator. When the altimeter is set, the pointers should indicate the local field elevation. If the altimeter registers within 75 feet of this elevation, it may be used, provided error is properly considered when the instrument is reset during flight.

9. Check operation of turn-and-bank indicator by observing proper response of needle and ball when turns are made during taxiing.

10. Check rate-of-climb indicator needle at zero.

Note

If needle does not indicate zero, tap instrument panel. If it still indicates incorrectly, have it adjusted.

11. Check accuracy of remote compass by comparing its reading to published runway heading.

12. Check carburetor ram-air control lever at RAM AIR.

13. Check instruments for readings within proper ranges.

14. Check operation of all radio equipment. Adjust tuning of required radio equipment as desired.

INSTRUMENT TAKE-OFF.

Preparation, power settings, take-off, and climb speeds are identical to those used in normal take-off. Use flaps about 20 degrees down for best obstacle clearance.

1. When cleared for take-off, taxi to center of runway and align airplane as nearly as possible straight down centerline of runway. Hold airplane with brakes and set directional gyro to published runway heading.

2. When ready, advance throttle smoothly and steadily to gate to obtain Take-off Power as quickly as possible and still maintain direction control against torque.

3. Do not attempt to lift tail too soon, as this increases torque action. Pushing the stick forward unlocks the tail wheel, thereby making steering difficult. The best take-off procedure is to hold the tail down until sufficient speed is attained for rudder control, and then to raise the tail slowly.

4. Maintain directional control by reference to directional gyro. Take off as airplane reaches normal VFR take-off airspeed.

TAKE-OFF SPEEDS

9,000 lb (no external load)	100 mph IAS
10,000 lb (external load)	105 mph IAS
11,000 lb (external load)	110 mph IAS

5. Raise gear as soon as altimeter and rate-of-climb indicator begin to register a climb.

6. Establish a normal climb.

7. Raise flaps when sufficient airspeed is attained and all obstacles are cleared. No sink is noticeable when the flaps are raised.

8. Reduce throttle setting and propeller control setting to climb at 2700 rpm and 46 in. Hg.

INSTRUMENT CLIMB.

1. Trim airplane at climbing speed of 170 mph.

2. Leave traffic and climb to altitude assigned. Do not exceed 30-degree angle of bank during climbing turns.

INSTRUMENT CRUISING FLIGHT.

No departure from normal cruise procedures is necessary. The most satisfactory cruising is obtained at 32 to 36 in. Hg at 2000 rpm. Adjust trim with caution so as not to overtrim. Maintain lateral trim by means of aileron trim tab for any unequal fuel distribution in main fuel tanks. Use fuel out of fuselage tank first until 25 gallons remains, so that CG is favorable for any situation that may arise. If flaps or landing gear are lowered, readjust power setting and retrim as necessary.

Note

To ensure the lowest fuel consumption on a long-range mission, the highest manifold pressure consistent with the Flight Operation Instruction Charts should be used with any given rpm setting. However, to minimize spark plug fouling resulting from prolonged cruising at low power (especially in the range from 1600 to 1900 rpm), high power (3000 rpm and 61 in. Hg) should be used continuously for one minute every 30 minutes when the fuel supply is adequate.

DESCENT.

Normal descent procedures are followed. Limit angle of bank to 30 degrees and single-needle-width turns.

CAUTION

Turn on defroster 10 or 15 minutes before descent, to avoid fogging of canopy and windshield.

HOLDING.

For holding with minimum fuel consumption, see Maximum Endurance Chart (figure A-9) for power settings at altitude where holding operation is being accomplished.

INSTRUMENT APPROACHES.

Use standard radio range letdown and low-visibility approaches. (See figure 9-1.)

GROUND-CONTROLLED APPROACH (GCA).

Procedure for landing under instrument conditions by use of directions from ground-controlled approach radar equipment after letdown on a radio range (figure 9-2) is as follows:

1. Establish contact with GCA over GCA pickup point.

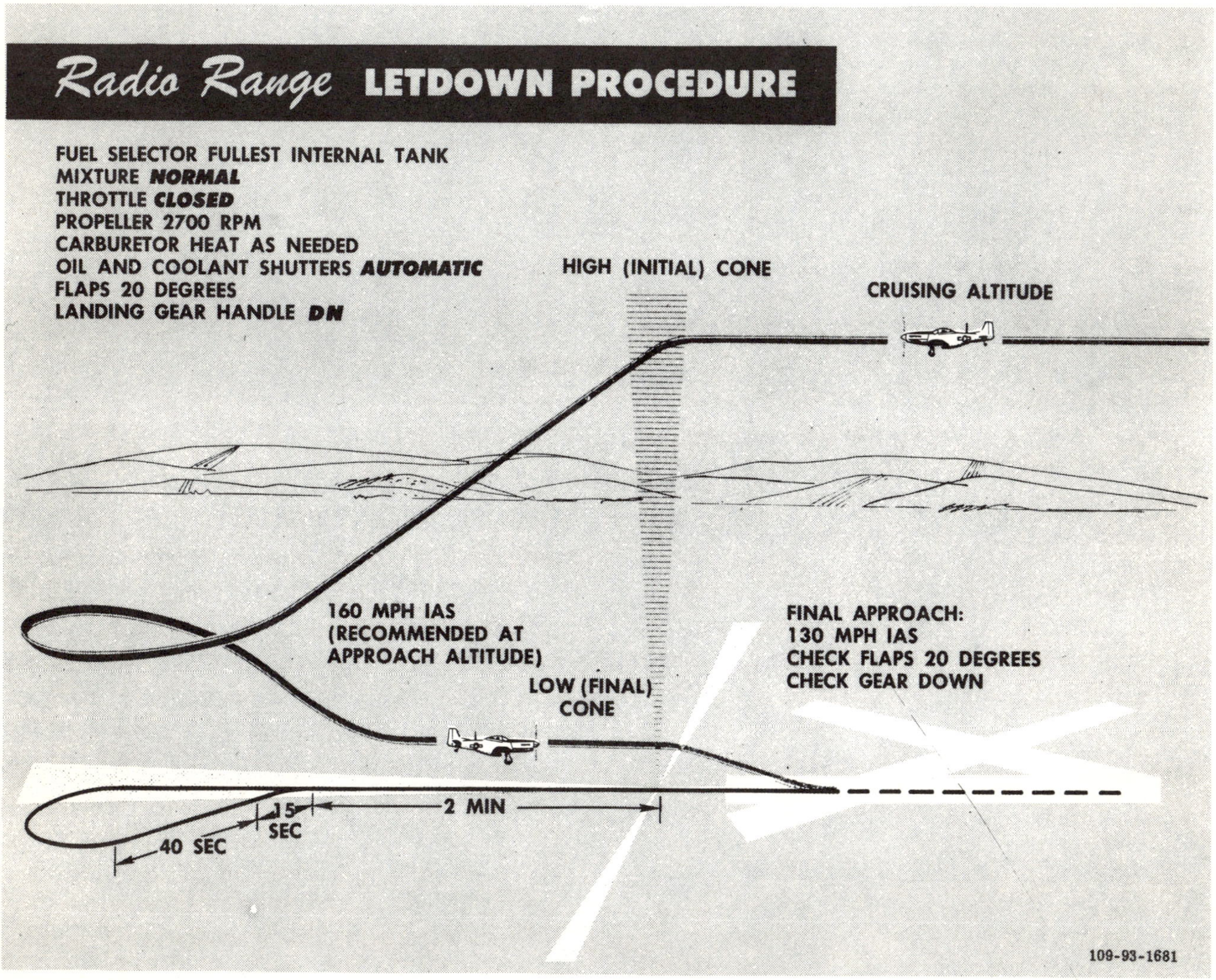

Figure 9-1

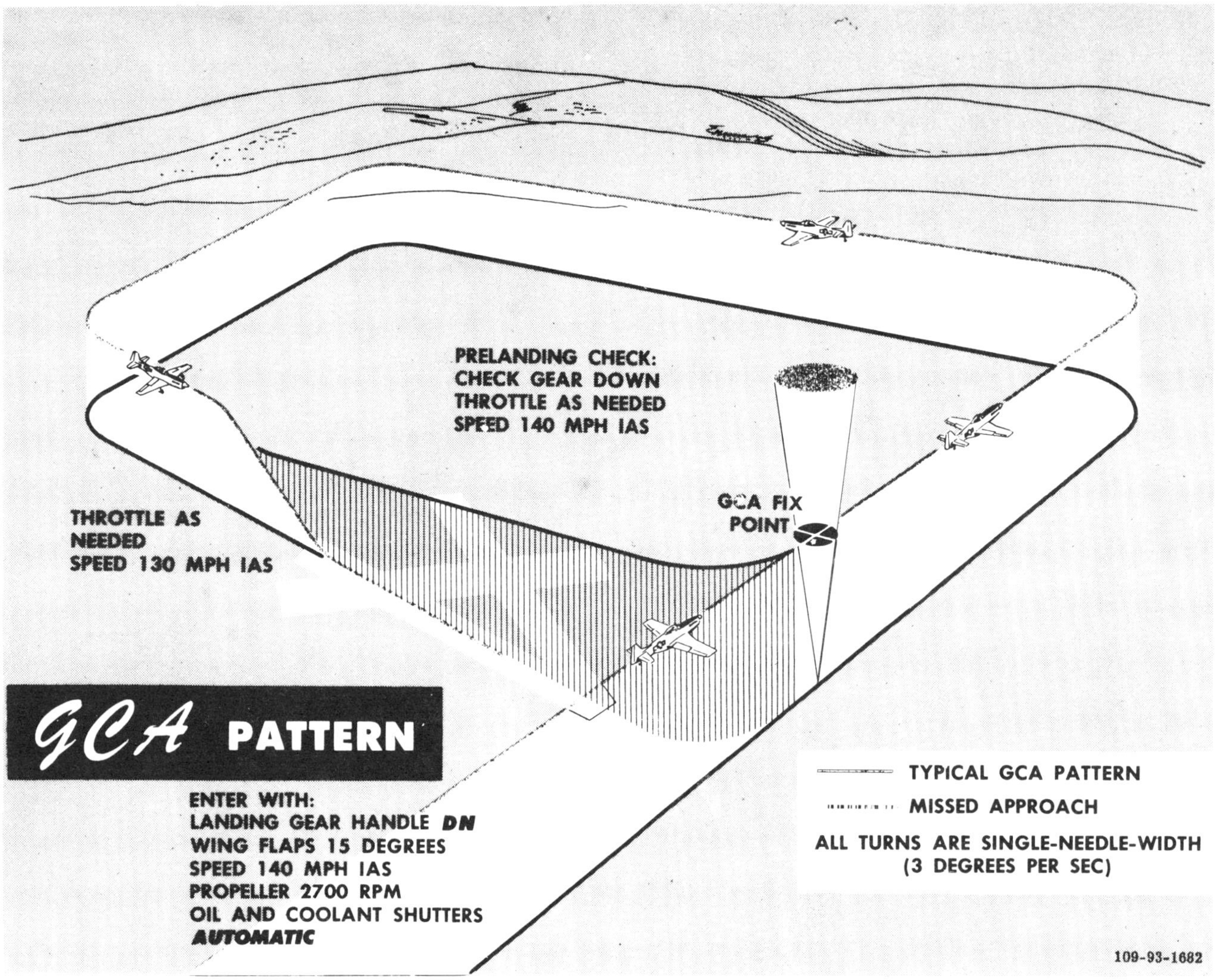

Figure 9-2

2. Hold 140 mph IAS until final turn is completed, running through GCA prelanding cockpit check as instructed by GCA controller.

3. After completing turn to final approach and before intercepting glide path, lower flaps 15 to 20 degrees, airspeed 140 mph.

4. As glide path is intercepted, reduce power setting to establish glide, and descend as directed by GCA final controller, using throttle as necessary.

MISSED-APPROACH GO-AROUND.

In case of missed approach, follow this procedure for go-around:

1. Open throttle smoothly to 45 in. Hg.
2. Maintain wings level, nose straight.
3. Landing gear up.
4. Raise flaps when at least 200 feet above ground and sufficient airspeed is reached.

ICING.

Ice normally adheres to the windshield, wing, stabilizer, vertical fin, and forward portions of the drop tanks. At the first sign of icing, change altitude immediately to get out of icing air layer. Ice accumulations increase drag and decrease lift, requiring an increase in power to maintain altitude and airspeed. To prevent carburetor ice during icing conditions, put carburetor ram-air control lever at UNRAMMED FILTERED AIR and, on late airplanes, carburetor hot-air control lever at HOT AIR. Use the two controls together. In case ice clogs the air duct, the hot-air door will open automatically. If ice accumulates on wings, make wide, shallow turns at a greater speed than normal, especially during approach. Use flaps with care. Remember, stalling speed increases with ice. Be sure pitot heater is on during icing conditions.

FLIGHT IN TURBULENCE AND THUNDERSTORMS.

CAUTION

Flight through a thunderstorm should be avoided if at all possible. Thunderstorm flying demands considerable instrument experience and should be intentionally undertaken only by well-qualified pilots. However, many routine flight operations require a certain amount of thunderstorm flying, since it is not always possible to avoid storm areas. At night, it is often impossible to detect individual storms and find the in-between clear areas.

Warning

A pilot using modern equipment and possessing a combination of proper experience, common sense, and instrument flying proficiency can safely fly thunderstorms.

Note

Normally, the least turbulent area in a thunderstorm is at altitudes of 6000 feet or less above the terrain. Altitudes between 10,000 and 20,000 feet are usually the most turbulent.

BEFORE TAKE-OFF.

Note the following precautions:

1. Make a thorough analysis of the general weather situation to determine thunderstorm areas, and prepare a flight plan which avoids thunderstorm areas whenever possible.

2. Be sure to check proper operation of all flight instruments, navigational equipment, pitot heater, carburetor air heat, and cockpit lighting before attempting flight through thunderstorm areas.

APPROACHING THE STORM.

It is imperative that you prepare the airplane before entering a zone of turbulent air. If a storm cannot be seen, its proximity can be detected by radio crash static. Prepare the airplane as follows:

1. Accurately fix position before actual entry into thunderstorm area.

2. Check 260 mph cruising speed power settings for comfortable penetration speed. (See figure 9-3.)

3. Mixture control NORMAL.

4. Pitot heater switch ON.

5. Carburetor ram-air and hot-air control levers adjusted as required.

6. Check suction gage for proper reading and gyro instruments for correct settings.

7. Turn off any radio equipment rendered useless by static.

8. Tighten safety belt. Lock shoulder harness.

9. At night, turn cockpit lights full bright, adjust seat low, and don't stare outside of airplane.

CAUTION

Do not lower landing gear or flaps, as they decrease the aerodynamic efficiency of the airplane.

IN THE STORM.

When in the thunderstorm, follow these procedures:

1. Throughout storm, maintain power settings and pitch attitude established before entering storm, unless airspeed falls off to 60 percent above power-on stalling speed or unless airspeed increases to approximately 30 percent above your penetration speed.

2. Devote all attention to flying airplane.

3. Expect turbulence, precipitation, and lightning. Don't allow these conditions to cause undue concern.

4. Maintain attitude. Concentrate principally on remaining level by reference to gyro horizon.

5. Maintain original heading. Do not make any turns unless absolutely necessary.

6. Don't chase airspeed indicator, since doing so will result in extreme airplane attitudes. If a sudden gust is encountered while the airplane is in a nose-high attitude, a stall might easily result. Because of rapid changes in vertical gust velocity or rain clogging the pitot tube, airspeed may momentarily fluctuate as much as 70 mph.

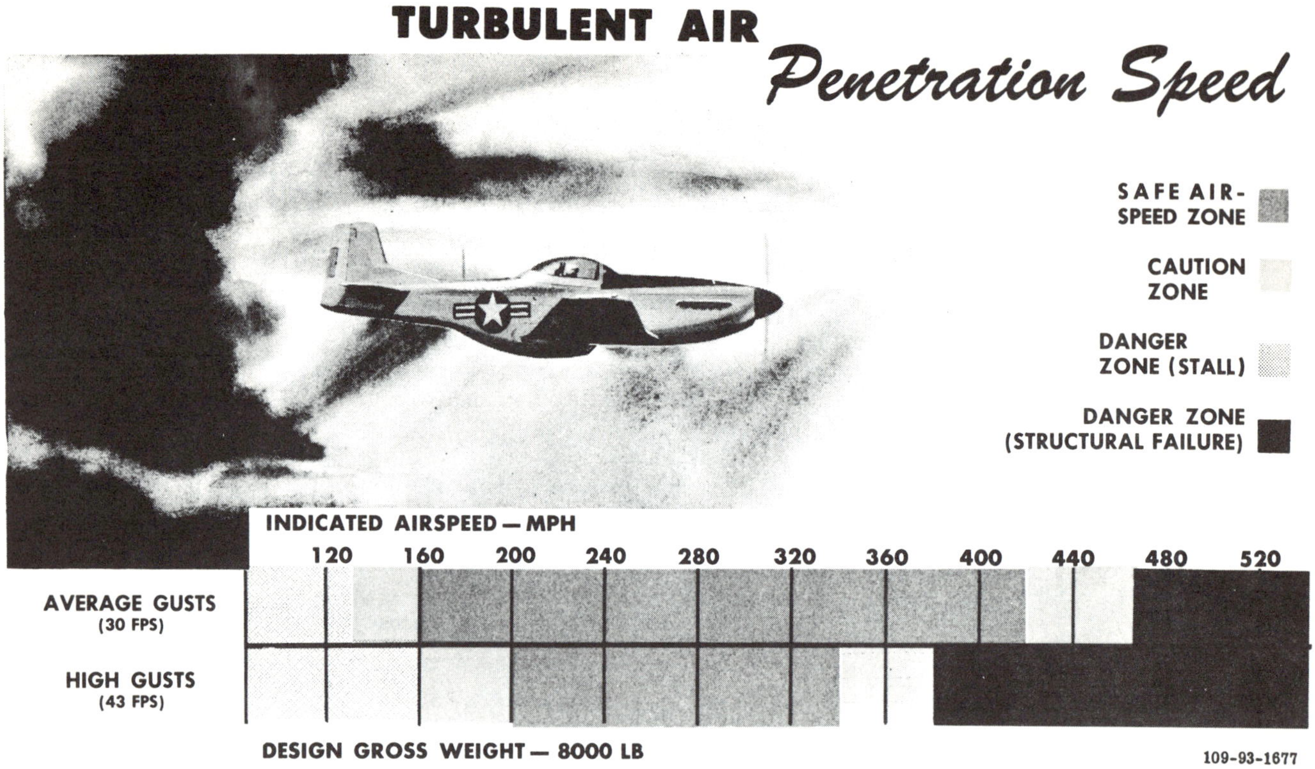

Figure 9-3

7. Use as little longitudinal control as possible to maintain your attitude, in order to minimize stresses imposed on airplane.

8. The altimeter may be unreliable in thunderstorms because of differential barometric pressure within the storm. A gain or loss of several thousand feet may be expected. Make allowance for this error in determining minimum safe altitude.

NIGHT FLYING

There are no predominant differences between night-flying procedures and day-flying procedures. Exhaust glare obviously is more pronounced during night flights, but should not be cause for alarm. Refer to Section II for night flight interior check and take-off and landing procedures.

COLD-WEATHER PROCEDURES

During cold-weather operation, normal operating procedures, as outlined in Section II, must be revised to include special inspection requirements and operating procedures necessitated by arctic conditions. Successful low-temperature operation is dependent upon the procedures that follow, especially those preparations made during engine shutdown and postflight servicing.

BEFORE ENTERING AIRPLANE.

1. Make thorough check of airplane for freedom from frost, snow, and ice. Include surfaces, controls, shock struts, hydraulic pistons, vents, breathers, etc. Make sure that all protective covers and excluder plugs have been removed.

2. Check that engine has been preheated in accordance with following chart:

PREHEAT CHART

Outside Air Temperature	Preheat Time (Minutes)
Above −18°C (0°F)	0
−18°C to −23°C (0°F to −10°F)	10
−23°C to −29°C (−10°F to −20°F)	20
−29°C to −34°C (−20°F to −30°F)	30
−34°C to −40°C (−30°F to −40°F)	40
−40°C to −46°C (−40°F to −50°F)	50
−46°C to −51°C (−50°F to −60°F)	60
−51°C to −54°C (−60°F to −65°F)	65

Note

The preheat times given in the chart are approximate and are based on preheating with a standard F-1A heater with one duct rerouted to the heater intake. Pull propeller through manually to determine need for additional preheat.

ON ENTERING AIRPLANE.

1. Check that cockpit, instrument panel, and windshield have been preheated when temperature is below –4°C (25°F).
2. Check controls and trim tabs for proper operation.
3. Make sure that all preheat equipment has been removed.
4. Make sure that an adequate auxiliary power cart (C-13A or equivalent) is connected.
5. Check that propeller can be pulled through manually and that there is fluid oil at "Y" drain immediately before attempting start.

STARTING ENGINE.

Make a normal start, following procedure outlined in Section II, as soon as possible after propeller is pulled through. More than normal priming is required at low temperatures during starting procedure and immediately after combustion until smooth engine operation is obtained. It is not considered harmful to prime continuously when necessary during the entire cranking period, but prime only when engine is turning over.

CAUTION

Do not open mixture control until engine is firing, to prevent excess fuel in induction system. If engine has not started after 2 minutes of cranking, disengage starter and allow starter to cool for one minute before making another attempt.

CAUTION

If there is no oil pressure after 30 seconds running, or if pressure drops to 0 after a few minutes of ground operation, stop engine immediately and investigate, to prevent excess wear and damage.

WARM-UP AND GROUND CHECK.

1. Move carburetor ram-air control lever to UNRAMMED FILTERED AIR and, on later airplanes, move carburetor hot-air control lever to HOT AIR after engine is started, to improve fuel vaporization and combustion and to reduce backfiring.
2. Do not increase engine speed above 1500 rpm until oil temperature rises to 20°C (68°F).
3. Ground-run engine for 30 minutes to remove excess fuel from oil if there is any possibility of overdilution.
4. Use firmly anchored wheel chocks for engine run-ups. Tie tail securely before attempting a full-power run-up.
5. Check wing flap operation.

TAXIING.

To preserve battery life, use only essential electrical equipment while taxiing at low engine speeds.

BEFORE TAKE-OFF.

1. Hold brakes and run up engine until spark plugs burn clean and engine is operating smoothly before checking magnetos.
2. Check flight controls for freedom of movement.
3. Use carburetor heat as required to keep carburetor air temperature within limits, to improve engine operation during take-off.
4. Turn pitot heater switch ON just before take-off.

TAKE-OFF.

At start of take-off run, advance throttle as rapidly as possible, to ensure that rated Take-off Power is obtainable.

WARNING

Discontinue take-off if required power is not available, because engine failure may occur.

AFTER TAKE-OFF.

1. After take-off from a wet-snow- or slush-covered runway, operate landing gear and flaps through several complete cycles to preclude their freezing.
2. Turn on gun and gun camera heaters.

ENGINE OPERATION IN FLIGHT.

On later airplanes, use carburetor heat as required to improve fuel vaporization and combat carburetor ice, but do not use carburetor heat above 12,000 feet, as resultant excessively lean mixtures will cause engine roughness due to the effect of heat on the altitude compensator of the carburetor.

Caution

Because of the constant-speed propeller and the automatic manifold pressure regulator, it is difficult to detect carburetor ice formation except by irregular engine operation.

OPERATION OF AIRPLANE SYSTEMS DURING FLIGHT.

1. Operate cockpit heat and defrosting system as required.

2. Increase propeller speed momentarily by approximately 200 rpm every half hour to ensure continued governing at extremely low temperatures. Return to desired cruising rpm as soon as tachometer indicates proper governing.

DESCENT.

Since temperature inversions occur frequently in the arctic, avoid engine overcooling during descents. Also, turn on windshield defrosting system to avoid fogging of the canopy before descent.

APPROACH.

1. Use carburetor heat when outside air temperatures are below −12°C (10°F).

2. Turn off all nonessential electrical equipment at least one minute before final approach, to reduce battery load when generator cuts out.

3. Pump brakes to chip away any accumulated ice.

STOPPING ENGINE.

1. Dilute engine in accordance with the following table for anticipated starting temperatures. Maintain oil temperature below 50°C (122°F), oil pressure above 15 psi, and 1300 to 1500 rpm during dilution period. Shut down engine with dilution switch engaged.

2. The following table gives dilution time for both standard dilution orifice (0.0625-inch diameter) and winterized orifice (0.111-inch diameter). The portion of the chart below the line (in excess of 10 percent dilution) is included for airplanes equipped with a Thompson centrifuge.

DILUTION TABLE

Temperature	Standard Minutes	Orifice (0.625 in.) Percent Dilution	Winterized Minutes	Orifice (0.111 in.) Percent Dilution
−12°C (10°F)	3		1.5	
−18°C (0°F)	4	10	2.0	10
−21°C (−5°F)	5		2.5	
−23°C (−10°F)	6		3.0	
−26°C (−15°F)	7		3.5	
−29°C (−20°F)	8	20	4.0	20
−32°C (−25°F)	9		4.5	
−34°C (−30°F)	10		5.0	
−37°C (−35°F)	11		5.5	
−40°C (−40°F)	12	30	6.0	30

Note

Do not dilute in excess of 10 percent unless a Thompson centrifuge is installed on the engine. Dilution over 10 percent will cause scavenge system failure and a dangerous loss of oil at high power settings.

3. Store unwinterized airplanes in a warm hangar if anticipated starting temperatures are below −18°C (0°F).

BEFORE LEAVING AIRPLANE.

1. Release brakes after wheels are chocked.

2. Leave canopy slightly open to allow air circulation within cockpit, to prevent canopy cracking from differential contraction and to decrease windshield and canopy frosting.

3. Whenever possible, leave airplane parked with full fuel tanks.

4. Have battery removed when airplane is parked outside at temperatures below −29°C (−20°F) for more than 4 hours or for any extended period of time.

HOT-WEATHER AND DESERT PROCEDURES

In general, hot-weather and desert procedures differ from normal procedures mainly in that additional precautions must be taken to protect the airplane from damage due to high temperatures and sand. Particular care should be taken to prevent the entrance of sand into the various airplane components and systems (engine, fuel system, pitot-static system, etc). All filters should be checked more often than under normal conditions. Units incorporating plastic and rubber parts should be protected as much as possible from excessive temperatures. Tires should be checked frequently for signs of blistering, etc.

BEFORE ENTERING AIRPLANE.

Check airplane for freedom from sand and dust (fungi in tropic climates). Include control hinges, hydraulic pistons, shock struts, etc, in this check. Remove protective covers and dust plugs.

ON ENTERING AIRPLANE.

1. Check control and trim tab operation for freedom of movement.
2. Check instruments and cockpit for freedom from sand and dust (fungi in tropic climates).

STARTING ENGINE.

1. Use normal starting procedure as outlined in Section II. Avoid overpriming.
2. Use filtered carburetor air for starting and ground operation as required.

WARM-UP AND GROUND CHECK.

Restrict ground operation to a minimum, to prevent overheating. Maintain a constant check on carburetor air and coolant temperature.

BEFORE TAKE-OFF.

Avoid take-off in a sand or dust storm whenever possible. Park airplane cross-wind and shut down engine.

TAKE-OFF.

1. Anticipate a longer take-off roll in high temperatures. (See Take-off Distances Chart, figure A-4.)
2. Check carburetor air and coolant temperatures closely during take-off.

APPROACH.

Switch to filtered carburetor air for landing.

ENGINE SHUTDOWN.

Shut down engine immediately on parking, to prevent overheating.

BEFORE LEAVING AIRPLANE.

1. Leave canopy partly open to permit air circulation within cockpit.
2. Make sure that protective covers and dust plugs are installed on engine, canopy, pitot tube, air ducts, and other parts as required.

Operating Data

APPENDIX I

TABLE OF CONTENTS. PAGE

INTRODUCTION.

There are two ways to perform a mission. The right way can be determined from the information presented in the charts on the following pages. If a pilot chooses to ignore the charts, he can fly any mission confident that the airplane is capable of greater performance than he is obtaining from it. These charts, which are easy to interpret, enable you to fly a greater distance at better cruising speed and arrive at your destination with more reserve fuel. A description of each chart and a sample problem to illustrate a typical training mission are also included.

AIRSPEED INSTALLATION AND COMPRESSIBILITY CORRECTION.

An Airspeed Installation Correction table (figure A-1) permits computing calibrated airspeed (CAS) from indicated airspeed (IAS). Indicated airspeed is the airspeed indicator reading. Calibrated airspeed is indicated airspeed corrected for installation error. An Airspeed Compressibility Correction table (figure A-1) permits computing equivalent airspeed (EAS) from calibrated airspeed (CAS). Equivalent airspeed (EAS) is calibrated airspeed corrected for compressibility error. True airspeed is equivalent airspeed corrected for atmospheric density.

AIRSPEED INSTALLATION CORRECTION

APPLY CORRECTION TO INSTRUMENT READING TO OBTAIN CALIBRATED AIRSPEED

GEAR AND FLAPS UP—CANOPY CLOSED		GEAR AND FLAPS DOWN—CANOPY OPEN	
MPH	CORRECTION	MPH	CORRECTION
100	5	80	6
120	5	90	3
140	4	100	1
160	4	110	-1
180	3	120	-2
200	3	130	-3
220	2	150	-4
240	2	170	-4
260	2		
280	2		
300	2		
350	3		
400	4		

AIRSPEED COMPRESSIBILITY CORRECTION

SUBTRACT CORRECTION FROM CALIBRATED AIRSPEED TO OBTAIN EQUIVALENT AIRSPEED

PRESSURE ALTITUDE	CALIBRATED AIRSPEED—MPH							
	150	200	250	300	350	400	450	500
5,000	0	0	1	1	2	3	4	5
10,000	0	1	2	3	4	6	8	
15,000	0	1	3	4	7	9		
20,000	1	2	4	6	10	14		
25,000	1	3	5	9	13			
30,000	2	4	7	12	18			
35,000	2	5	10	16				

109-93-1734

Figure A-1

FREE AIR TEMPERATURE *Correction Chart*

ALTITUDE—FEET	INDICATED AIRSPEED—MPH				
	150	200	250	300	320
0		4	6	8	9
5,000		4	7	9	11
10,000		5	8	11	12
15,000	4	6	9	13	15
20,000	5	8	11	16	18
25,000	6	9	13	19	21
30,000	7	11	17	22	24
35,000	8	13	20		
40,000	10	17	24		
45,000	12	21			

SUBTRACT CORRECTION SHOWN FROM CARBURETOR AIR TEMPERATURE TO OBTAIN FREE AIR TEMPERATURE IN DEGREES CENTIGRADE.

DATA BASIS: FLIGHT TEST
DATA AS OF: 9-2-53

109-93-1745

Figure A-2

FREE AIR TEMPERATURE CORRECTION.

Since no free air temperature gage is provided in this airplane, a chart for converting indicated carburetor air temperature to free air temperature is given in figure A-2. The corrected free air temperature can be used with calibrated airspeed to obtain true airspeed.

EXAMPLE—USE OF CORRECTION TABLES.

An airplane is flying at 25,000 feet pressure altitude. The indicated carburetor air temperature is −15°C, and the indicated airspeed reading is 300 mph. What is the true airspeed?

Airspeed indicator reading (IAS)	300 mph
Correction for installation error	2 mph
Calibrated airspeed (CAS)	302 mph

Indicated carburetor air temperature	−15°C
Correction to obtain free air temperature	−19°C
Corrected free air temperature	−34.0°C

Use these values of CAS and free air temperature with a Type D-4 or Type G-1 airspeed computer to determine the true airspeed of 438 mph.

When a Type AN5835-1 dead-reckoning computer is used, CAS usually must be corrected for compressibility error.

Calibrated airspeed (CAS)	302 mph
Compressibility Error	−9 mph
Equivalent Airspeed (EAS)	293 mph

Use this value of EAS with the dead-reckoning computer to determine the true airspeed of 438 mph.

TAKE-OFF DISTANCES.

The Take-off Distances charts (figure A-4) give take-off ground-run distances and total distances to clear a 50-foot obstacle. The charts are tabulated for several different gross weights, altitudes, and temperatures on a hard-surface runway. Distances given are for standard flaps-up take-offs. For a minimum-run take-off, refer to Section II.

CLIMB.

Best climb speed, fuel consumption, time to climb, and rate of climb (using Military Power or Normal Power) can be determined for different configurations from the Military Power and Normal Power Climb charts (figures A-5 and A-6). A fuel allowance for warm-up, taxi, and take-off is listed in the column labeled "SEA

LEVEL." Fuel requirements listed at other altitudes include this allowance plus the fuel required to climb from sea level. Fuel required for an in-flight climb from one altitude to another is the difference between the tabulated fuel required to climb to each altitude from sea level.

LANDING DISTANCES.

The Landing Distances chart (figure A-7) shows the distances required for ground roll and for landing over a 50-foot obstacle. Distances for landings on a hard-surface runway are furnished for several altitudes and gross weights. Best speeds are shown for power-off approach. Distances given are airplane requirements under normal service conditions with no wind and with flaps full down.

MAXIMUM ENDURANCE.

Airspeeds, power settings, and fuel flow rates for maximum endurance flight are shown in the Maximum Endurance charts (figure A-9) for several configurations and altitudes. The Maximum Endurance charts give the power settings and fuel flows for maximum *time* in the air and should not be confused with the "MAXIMUM AIR RANGE" section of the Flight Operation Instruction Charts, in which the power setting and fuel flows are for maximum *distance,* not maximum *time.*

COMBAT ALLOWANCE.

The Combat Allowance chart (figure A-8) shows the variation with altitude in manifold pressure and fuel flow at Take-off Power (Military Power).

FLIGHT OPERATION INSTRUCTION CHARTS.

To assist in selecting the engine operating conditions required for obtaining various ranges, Flight Operation Instruction Charts (figures A-10 through A-16) are provided. Each chart is divided into five main columns. Data listed under Column I is for emergency high-speed cruising at Maximum Continuous Power. Operating conditions in Columns II, III, IV, and V give progressively greater ranges at lower cruising speeds. Ranges shown in any column for a given fuel quantity can be obtained at various altitudes by use of the power settings listed in the lower half of the chart in the same column. The speeds quoted on the chart are those obtained with gross weight equal to the high limit of the chart weight band. Speeds are shown to the nearest 5 mph. No allowances are made for wind, navigational error, simulated combat, formation flights, etc; therefore, such allowances must be made as required.

USE OF CHARTS.

To use the charts, first select the Flight Operation Instruction Chart applicable to your flight plan, determined in this airplane by gross weight at take-off and by external load. Then enter the chart at a fuel quantity equal to, or less than, the total amount in the airplane minus all allowances. (Ranges listed for each fuel quantity are based on use of the entire quantity in level flight when cruising at the recommended operating conditions.) Fuel allowance for warm-up, taxi, take-off, and climb is obtained from the desired climb chart (figure A-5 or A-6). Other allowances based on the type of mission, terrain over which the flight is to be made, and weather conditions are dictated by local policy. If your flight plan calls for a continuous flight at reasonably constant cruising power, compute the fuel required and flight time as for a single-section flight. Otherwise, the flight must be broken up into sections and each leg of the flight planned separately. The flight plan may be changed at any time en route, and the chart will show the balance of range available at various cruising powers and altitudes if the instructions printed at the top of the chart are followed.

SAMPLE PROBLEMS.

PROBLEM 1.

A bombing run must be made on a target 231 statute miles from the home field. A secondary target, 70 statute miles from the bomb target and 275 miles from the home field, is to be strafed to lend ground support. Military Power will be used during the runs on both target areas. The bomb run will be initiated from 5000 feet altitude, the gunnery runs will be made at sea level plus 50 feet, and run-in to the bomb target will be made "on the deck" (sea level plus 50 feet) to avoid radar detection. The run to the secondary target will be made "on the deck" as well. Maximum Continuous Power will be used on both of these legs. After completion of the gunnery runs, a climb from sea level to 10,000 feet will be made on course to the home field. Cruise back will be at 10,000 feet. (See figure A-3.)

Write down the conditions of the problem:

Required range ....576 statute miles
Weather ....CAVU
Wind ....0 mph on all legs
Airplane basic weight......7135 pounds (includes trapped fuel and oil, and miscellaneous equipment)
Crew weight (one) ....230 pounds
Oil (12.5 gal) ....93 pounds
Maximum internal fuel (245 gal) ....1470 pounds

K-14B gun sight .... 140 pounds
Armament .... 1350 pounds (includes 1000 rounds ammunition, gun camera, two 500-pound general-purpose bombs, and items necessary for installation)
Total gross weight......... 10,418 pounds

Now that the conditions of the flight are determined, it is necessary to establish a flight plan. Since the charts give only cruise ranges under no-wind conditions and do not include any reserves, it is necessary first to compute all allowances and reserves that will be required to cover warm-up, take-off, climb, Military Power operation, and any unexpected difficulties. Determine fuel available for cruise flight by deducting necessary fuel allowances and reserves from actual fuel aboard as follows:

General reserve for unexpected difficulties .... 42 gallons

Note in Column V of figure A-10 that at 5000 feet, 42 gallons of fuel represents one hour's flying time. A one-hour fuel reserve is considered sufficient for this mission.

Warm-up, take-off, and climb to 50 feet .... 15 gallons

The Normal Power Climb chart (figure A-5) shows that 15 gallons is required for warm-up, take-off, and climb to 50 feet.

Military Power allowance .... 28 gallons

This figure is obtained by multiplying the Military Power fuel consumption at sea level (given in the Combat Allowance chart, figure A-8) by the total time spent at this power; i.e., 5 minutes on bomb target (5 minutes X 3.0 gpm = 15 gallons) + 5 minutes ground support (5 minutes X 2.5 gpm = 12.5 gallons) = 28 gallons.

Climb from sea level to 5000 feet........ 7 gallons

The Normal Power Climb chart (figure A-5) shows that 22 gallons is required to climb to 5000 feet, less 15 gallons warm-up and take-off allowance, or 7 gallons (22 − 15 = 7 gallons). Observe that a distance of 13 statute miles is covered during the climb to bomb-run altitude. Therefore, the climb to bombing altitude should be started 13 miles out from the target for arrival over the target at the proper altitude.

Descent to sea level from 5000 feet...... 0 gallons

The descent from bombing altitude to sea level plus 50 feet (the altitude used for run-in on ground support target) is considered to be included in the fuel used during the bomb run at Military Power.

Climb from sea level to 10,000 feet.... 10 gallons

After the gunnery runs are completed, the airplane is flown to 10,000 feet on course to the home field. The Normal Power Climb chart (figure A-5) shows that 25 gallons is required to climb to 10,000 feet, less 15 gallons warm-up and take-off allowance, or 10 gallons (25 − 15 = 10 gallons). During the climb, a distance of 22 statute miles is covered.

Collecting all the required fuel allowances:

General reserve for unexpected difficulties .... 42 gallons
Warm-up, take-off, and climb to 50 feet .... 15 gallons
Military Power allowance .... 28 gallons
Climb from sea level to 5000 feet.... 7 gallons
Descent to sea level from 5000 feet .... 0 gallons
Climb from sea level to 10,000 feet .... 10 gallons
Total fuel allowance .... 102 gallons

Therefore, the actual fuel available for cruising is 143 gallons (245 − 102 = 143 gallons). In the climb from sea level to 5000 feet, a total of 13 statute miles was covered so that the total range, on the first leg, to be flown with Maximum Continuous Power is 218 statute miles (231 − 13 = 218 miles). By reference to figures A-11 and A-12, the fuel required can be determined from Column I, at sea level (Maximum Continuous Power operation). Range divided by true airspeed, then multiplied by fuel flow, gives fuel required; i.e., 218 miles ÷ 292 mph = 0.746 hour, and 0.746 hour X 86 gph = 64 gallons. This leaves 79 gallons for the remaining two legs (143 − 64 = 79 gallons). The second leg is figured the same as the first, using figure A-10; remember, the bombs were disposed of, at the end of the first leg. Column I (figure A-10) shows a true airspeed of 315 mph with a fuel flow of 86 gph. The fuel required for the second leg of 70 statute miles is 19 gallons (70 ÷ 315 = 0.222 hour, and 0.222 hour X 86 gph = 19 gallons). This leaves 60 gallons (79 − 19 = 60 gallons) for the homeward-bound leg. Since 22 statute miles of this leg is covered climbing to 10,000 feet, this leaves a distance of 253 statute miles to be flown at 10,000 feet with the wing rack configuration. By reference to figure A-10, it can be seen that the power settings in Columns III, IV, or V will give the required mileage.

Going from Column III to Column V gives a progressive increase in range at a sacrifice in speed, as well as an added reserve. Suppose Column III is picked; the fuel required will be 55 gallons (253 miles ÷ 325 mph = 0.778 hour, and 0.778 hour X 71 gph = 55 gallons). This gives a 5-gallon surplus (60 − 55 = 5 gallons) which, if added to the original reserve quantity, gives a total reserve of 53 gallons (48 + 5 = 53 gallons). This, then, is a quick solution to the problem.

PROBLEM 2.

Suppose that the estimate of 5 minutes of Military Power at each of the targets was too low, and the actual time spent was 10 minutes per target. Therefore, the original 28-gallon Military Power allowance must be increased to 55 gallons [(10 minutes X 3.0 gpm = 30 gallons) + (10 minutes X 2.5 gpm = 25 gallons) = 55 gallons], and consideration of the remainder of the mission must be made during flight. If the remaining leg of the mission is flown as originally planned (Column III, figure A-10), the additional Military Power allowance may be subtracted from the allowed reserve of 55 gallons, leaving a reserve at the end of the mission of 28 gallons. However, if a greater reserve is desired, the last leg of the mission may be flown at slightly lower power settings and speeds, such as those listed in Column IV of figure A-10. Note in Column IV that the remaining 253 statute miles of cruising requires only 46 gallons of fuel. This compares with the 55 gallons required to travel the same distance using Column III power settings. The net savings in fuel, by using Column IV instead of Column III, is 9 gallons (55 — 46 = 9 gallons), which at 5000 feet represents an additional 0.191 hour at maximum range or the equivalent of 50 additional statute miles (9 gallons ÷ 47 gph = 0.191 hour, and 0.191 hour X 261 mph = 50 statute miles).

If for some reason the 55-gallon reserve is to be considered for a holding or orbiting procedure where *time* in the air is important rather than range, consult figure A-9 to determine that the 55-gallon reserve represents 1.48 hours of flying time at 5000 feet (55 gallons ÷ 37 gph = 1.48 hours).

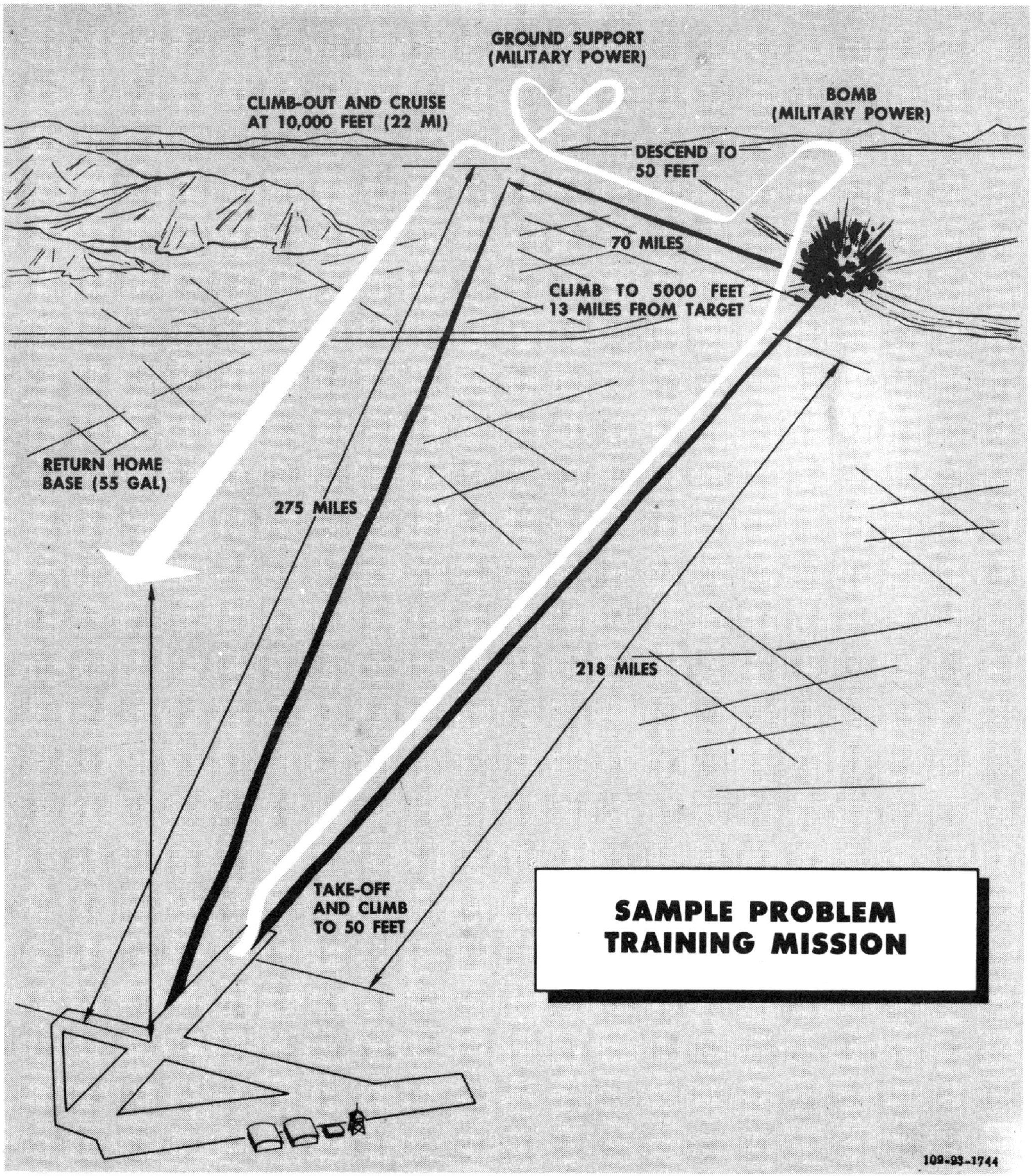
GROUND SUPPORT
(MILITARY POWER)
CLIMB-OUT AND CRUISE
AT 10,000 FEET (22 MI)
BOMB
(MILITARY POWER)
DESCEND TO
50 FEET
70 MILES
CLIMB TO 5000 FEET
13 MILES FROM TARGET
RETURN HOME
BASE (55 GAL)
275 MILES
218 MILES
TAKE-OFF
AND CLIMB
TO 50 FEET
SAMPLE PROBLEM
TRAINING MISSION
109-93-1744

Figure A-3

WADC
Form 241G
(11 Jun 51)

TAKE-OFF DISTANCES

(FEET)

HARD-SURFACE RUNWAY

MODEL: F-51D

ENGINE (S): (1) V-1650-7

GROSS WEIGHT	PRESSURE ALTITUDE	−5 DEGREES CENTIGRADE				+15 DEGREES CENTIGRADE				+35 DEGREES CENTIGRADE				+55 DEGREES CENTIGRADE			
		ZERO WIND		30-KNOT WIND		ZERO WIND		30-KNOT WIND		ZERO WIND		30-KNOT WIND		ZERO WIND		30-KNOT WIND	
		GROUND RUN	TO CLEAR 50 FT OBST.	GROUND RUN	TO CLEAR 50 FT OBST.	GROUND RUN	TO CLEAR 50 FT OBST.	GROUND RUN	TO CLEAR 50 FT OBST.	GROUND RUN	TO CLEAR 50 FT OBST.	GROUND RUN	TO CLEAR 50 FT OBST.	GROUND RUN	TO CLEAR 50 FT OBST.	GROUND RUN	TO CLEAR 50 FT OBST.
13,000 LB	SL	2050	3050	1050	1700	2350	3450	1250	2000	2800	3900	1550	2350	3250	4550	1850	2750
	1000	2200	3250	1150	1850	2600	3700	1400	2200	3050	4250	1700	2550	3550	4950	2050	3000
	2000	2400	3500	1300	2000	2800	4000	1550	2350	3300	4600	1900	2800	3900	5350	2250	3250
	3000	2600	3750	1400	2200	3050	4300	1700	2550	3600	5000	2100	3050	4250	5800	2500	3600
	4000	2800	4000	1550	2400	3350	4650	1900	2800	3950	5400	2300	3300	4700	6300	2800	3950
	5000	3100	4350	1750	2600	3650	5050	2100	3100	4350	5900	2600	3700	5200	6900	3150	4400
12,000 LB	SL	1650	2550	850	1400	1950	2950	1000	1650	2300	3350	1200	1900	2650	3800	1400	2200
	1000	1800	2750	900	1550	2150	3200	1100	1800	2500	3600	1350	2100	2850	4100	1600	2400
	2000	1950	2950	1000	1650	2300	3400	1250	1950	2700	3900	1500	2300	3100	4400	1750	2650
	3000	2150	3200	1100	1800	2500	3650	1350	2150	2950	4200	1650	2500	3400	4750	1950	2900
	4000	2350	3400	1250	1950	2750	3900	1500	2300	3200	4500	1800	2700	3750	5150	2200	3150
	5000	2550	3650	1350	2150	3000	4200	1650	2500	3500	4850	2000	2950	4100	5600	2400	3450
11,000 LB	SL	1350	2200	650	1200	1600	2500	800	1350	1850	2850	950	1600	2150	3200	1150	1850
	1000	1450	2300	700	1250	1750	2650	850	1500	2000	3050	1050	1700	2350	3400	1250	2050
	2000	1600	2500	800	1350	1850	2850	950	1600	2200	3250	1150	1850	2550	3700	1400	2150
	3000	1700	2650	850	1450	2000	3050	1050	1750	2400	3500	1300	2000	2800	3950	1550	2350
	4000	1850	2850	950	1600	2200	3250	1150	1900	2600	3750	1400	2200	3050	4250	1700	2550
	5000	2050	3050	1050	1750	2400	3500	1300	2050	2850	4050	1600	2400	3300	4600	1900	2800
10,000 LB	SL	1050	1850	500	1000	1250	2100	600	1150	1450	2350	750	1300	1700	2650	850	1450
	1000	1150	2000	550	1050	1400	2250	700	1200	1600	2500	800	1400	1850	2850	950	1600
	2000	1250	2100	600	1150	1500	2400	750	1300	1750	2700	900	1500	2000	3050	1050	1750
	3000	1400	2250	650	1200	1650	2500	800	1400	1900	2900	1000	1650	2200	3250	1150	1900
	4000	1500	2400	750	1300	1750	2700	900	1550	2100	3100	1100	1800	2400	3500	1300	2050
	5000	1650	2550	800	1450	1900	2900	1000	1650	2300	3300	1200	1950	2650	3800	1450	2250

REMARKS:
1. Take-off distances are aircraft requirements under normal service conditions.
2. Take-off Power, 3000 rpm 61 in. Hg.
3. Flaps up.

DATA AS OF 8-1-53
BASED ON FLIGHT TEST

109-93-1772

FUEL GRADE: 100/130
FUEL DENSITY: 6.0 LB/GAL

Figure A-4. Take-off Distances (Sheet 1 of 2)

WADC
Form 241G
(11 Jun 51)

TAKE-OFF DISTANCES

(FEET)

HARD-SURFACE RUNWAY

MODEL: F-51D

ENGINE (S): (1) V-1650-7

GROSS WEIGHT	PRESSURE ALTITUDE	−5 DEGREES CENTIGRADE				+15 DEGREES CENTIGRADE				+35 DEGREES CENTIGRADE				+55 DEGREES CENTIGRADE			
		ZERO WIND		30-KNOT WIND		ZERO WIND		30-KNOT WIND		ZERO WIND		30-KNOT WIND		ZERO WIND		30-KNOT WIND	
		GROUND RUN	TO CLEAR 50 FT OBST.	GROUND RUN	TO CLEAR 50 FT OBST.	GROUND RUN	TO CLEAR 50 FT OBST.	GROUND RUN	TO CLEAR 50 FT OBST.	GROUND RUN	TO CLEAR 50 FT OBST.	GROUND RUN	TO CLEAR 50 FT OBST.	GROUND RUN	TO CLEAR 50 FT OBST.	GROUND RUN	TO CLEAR 50 FT OBST.
9000 LB	SL	850	1550	350	800	1000	1750	450	900	1150	1950	550	1050	1350	2150	650	1150
	1000	950	1650	400	850	1100	1850	500	950	1250	2050	600	1100	1450	2300	700	1250
	2000	1000	1750	450	900	1150	1950	550	1050	1350	2200	650	1200	1550	2450	800	1350
	3000	1100	1850	500	950	1250	2050	600	1100	1450	2350	750	1300	1700	2650	850	1450
	4000	1150	1950	550	1050	1350	2200	650	1200	1600	2500	800	1400	1850	2800	950	1600
	5000	1250	2100	600	1150	1500	2350	750	1300	1750	2650	900	1500	2000	3000	1050	1750
	SL																
	1000																
	2000																
	3000																
	4000																
	5000																
	SL																
	1000																
	2000																
	3000																
	4000																
	5000																
	SL																
	1000																
	2000																
	3000																
	4000																
	5000																

REMARKS:
1. Take-off distances are aircraft requirements under normal service conditions.
2. Take-off Power, 3000 rpm 61 in. Hg.
3. Flaps up.

DATA AS OF 8-1-53
BASED ON FLIGHT TEST

109-93-1773

FUEL GRADE: 100/130
FUEL DENSITY: 6.0 LB/GAL

Figure A-4. Take-off Distances (Sheet 2 of 2)

WADC
Form 2411
(11 Jun 51)

NORMAL POWER CLIMB CHART

STANDARD DAY

MODEL: F-51D

ENGINE(S): (1) V-1650-7

CONFIGURATION (left): SIX 5 IN. ROCKETS - PLUS TWO 75 GAL TANKS, TWO 110 GAL TANKS, TWO 1000 LB BOMBS, OR ONE 1000 LB BOMB PLUS ONE 110 GAL TANK
GROSS WEIGHT: 13,000 TO 11,000 POUNDS

CONFIGURATION (right): SIX 5 IN. ROCKETS - PLUS TWO 75 GAL TANKS, TWO 110 GAL TANKS, TWO 1000 LB BOMBS, OR ONE 1000 LB BOMB PLUS ONE 110 GAL TANK
GROSS WEIGHT: 11,000 POUNDS OR LESS

APPROXIMATE				MP	CAS	PRESSURE ALTITUDE	CAS	MP	APPROXIMATE			
RATE OF CLIMB	FROM SEA LEVEL			(IN. Hg)	(MPH)	(FEET)	(MPH)	(IN. Hg)	FROM SEA LEVEL			RATE OF CLIMB
	DISTANCE	TIME	FUEL	(2)				(2)	FUEL	TIME	DISTANCE	
700	0	0	15(1)	46	175	SEA LEVEL	175	46	15(1)	0	0	1000
700	23	7	26	46	185	5,000	185	46	23	5	17	950
650	49	15	38	46	190	10,000	190	46	32	11	36	950
600	80	23	50	46	190	15,000	190	46	41	16	58	900
300	123	33	67	F.T.	185	20,000	185	F.T.	52	24	87	850
200	214	54	99	46	180	25,000	180	46	68	34	133	300
						30,000	175	F.T.	93	50	204	350
						35,000						
						40,000						
						45,000						

CONFIGURATION (left): TWO 1000 LB BOMBS, TWO 110 GAL TANKS, TEN 5 IN. ROCKETS, OR ONE 1000 LB BOMB PLUS ONE 110 GAL TANK
GROSS WEIGHT: 12,200 TO 10,300 POUNDS

CONFIGURATION (right): TWO 1000 LB BOMBS, TWO 110 GAL TANKS, TEN 5 IN. ROCKETS, OR ONE 1000 LB BOMB PLUS ONE 110 GAL TANK
GROSS WEIGHT: 10,300 POUNDS OR LESS

APPROXIMATE				MP	CAS	PRESSURE ALTITUDE	CAS	MP	APPROXIMATE			
RATE OF CLIMB	FROM SEA LEVEL			(IN. Hg)	(MPH)	(FEET)	(MPH)	(IN. Hg)	FROM SEA LEVEL			RATE OF CLIMB
	DISTANCE	TIME	FUEL	(2)				(2)	FUEL	TIME	DISTANCE	
900	0	0	15(1)	46	175	SEA LEVEL	175	46	15(1)	0	0	1250
900	18	6	24	46	185	5,000	185	46	21	4	13	1250
900	38	11	32	46	190	10,000	190	46	27	8	27	1250
850	60	17	42	46	190	15,000	190	46	34	12	42	1250
550	90	24	53	F.T.	185	20,000	185	F.T.	41	17	62	850
400	140	36	71	46	180	25,000	180	46	51	23	90	750
250	208	51	94	F.T.	175	30,000	175	F.T.	63	31	125	550
						35,000	165	F.T.	83	45	192	150
						40,000						
						45,000						

REMARKS:
1. Warm-up, taxi, and take-off allowance.
2. 2700 rpm.
3. Blower shift automatic.

LEGEND

RATE OF CLIMB - FEET PER MINUTE
DISTANCE - STATUTE MILES
TIME - MINUTES
FUEL - US. GALLONS
MP - MANIFOLD PRESSURE
CAS - CALIBRATED AIRSPEED
F.T. - FULL THROTTLE

DATA AS OF 8-1-53
BASED ON FLIGHT TEST

FUEL GRADE: 100/130
FUEL DENSITY: 6.0 LB/GAL

109-93-1777

Figure A-5. Normal Power Climb (Sheet 1 of 2)

WADC
Form 2411
(11 Jun 51)

NORMAL POWER CLIMB CHART

STANDARD DAY

MODEL: F-51D ENGINE(S): (1) V-1650-7

CONFIGURATION: TWO 500 LB BOMBS OR TWO 75 GAL TANKS — GROSS WEIGHT: 11,200 TO 9800 POUNDS						PRESSURE ALTITUDE (FEET)	CONFIGURATION: TWO 500 LB BOMBS OR TWO 75 GAL TANKS — GROSS WEIGHT: 9800 POUNDS OR LESS					
RATE OF CLIMB	APPROXIMATE FROM SEA LEVEL DISTANCE	TIME	FUEL	MP (IN. Hg) (2)	CAS (MPH)		CAS (MPH)	MP (IN. Hg) (2)	APPROXIMATE FROM SEA LEVEL FUEL	TIME	DISTANCE	RATE OF CLIMB
1150	0	0	15[1]	46	175	SEA LEVEL	175	46	15[1]	0	0	1500
1200	13	4	22	46	185	5,000	185	46	20	3	11	1500
1200	28	8	28	46	190	10,000	190	46	25	7	22	1550
1200	44	13	35	46	190	15,000	190	46	31	10	35	1550
900	64	17	42	F.T.	185	20,000	185	F.T.	36	14	50	1200
800	91	24	52	46	180	25,000	180	46	43	18	69	1050
550	124	31	64	F.T.	175	30,000	175	F.T.	51	23	93	850
150	194	46	84	F.T.	165	35,000	165	F.T.	62	31	129	450
				F.T.	155	40,000						
						45,000						

CONFIGURATION: WING RACKS — GROSS WEIGHT: 10,200 POUNDS OR LESS						PRESSURE ALTITUDE (FEET)	CONFIGURATION: — GROSS WEIGHT:					
RATE OF CLIMB	APPROXIMATE FROM SEA LEVEL DISTANCE	TIME	FUEL	MP (IN. Hg) (2)	CAS (MPH)		CAS (MPH)	MP (IN. Hg)	APPROXIMATE FROM SEA LEVEL FUEL	TIME	DISTANCE	RATE OF CLIMB
1500	0	0	15[1]	46	175	SEA LEVEL						
1550	10	3	20	46	185	5,000						
1550	22	6	25	46	190	10,000						
1600	34	10	30	46	190	15,000						
1300	48	13	36	F.T.	185	20,000						
1150	66	17	42	46	180	25,000						
900	88	22	50	F.T.	175	30,000						
500	121	29	59	F.T.	165	35,000						
				F.T.	155	40,000						
						45,000						

REMARKS:
1. Warm-up, taxi, and take-off allowance.
2. 2700 rpm.
3. Blower shift automatic.

LEGEND

RATE OF CLIMB - FEET PER MINUTE
DISTANCE - STATUTE MILES
TIME - MINUTES
FUEL - US. GALLONS
MP - MANIFOLD PRESSURE
CAS - CALIBRATED AIRSPEED
F.T. - FULL THROTTLE

DATA AS OF 8-1-53
BASED ON FLIGHT TEST

FUEL GRADE: 100/130
FUEL DENSITY: 6.0 LB/GAL

109-93-1778

Figure A-5. Normal Power Climb (Sheet 2 of 2)

WADC
Form 2411
(11 Jun 51)

MILITARY POWER CLIMB CHART

STANDARD DAY

MODEL: F-51D ENGINE(S): (1) V-1650-7

CONFIGURATION: SIX 5 IN. ROCKETS - PLUS TWO 75 GAL TANKS, TWO 110 GAL TANKS, TWO 1000 LB BOMBS, OR ONE 1000 LB BOMB PLUS ONE 110 GAL TANK — GROSS WEIGHT: 13,000 TO 11,000 POUNDS							CONFIGURATION: SIX 5 IN. ROCKETS - PLUS TWO 75 GAL TANKS, TWO 110 GAL TANKS, TWO 1000 LB BOMBS, OR ONE 1000 LB BOMB PLUS ONE 110 GAL TANK — GROSS WEIGHT: 11,000 POUNDS OR LESS					
APPROXIMATE				MP (IN. Hg) (2)	CAS (MPH)	PRESSURE ALTITUDE (FEET)	CAS (MPH)	MP (IN. Hg) (2)	APPROXIMATE			
RATE OF CLIMB	FROM SEA LEVEL DISTANCE	TIME	FUEL						FROM SEA LEVEL FUEL	TIME	DISTANCE	RATE OF CLIMB
1350	0	0	15(1)	61	175	SEA LEVEL	175	61	15(1)	0	0	1750
1250	12	4	25	61	185	5,000	185	61	23	3	9	1650
1150	27	8	36	61	190	10,000	190	61	31	6	20	1600
850	45	13	80	F.T.	190	15,000	190	F.T.	41	9	33	1250
550	78	21	102	61	185	20,000	185	61	55	14	54	900
450	120	30	130	61	180	25,000	180	61	71	20	79	800
150	198	48	174	F.T.	175	30,000	175	F.T.	91	28	114	500
					165	35,000	165	F.T.	122	44	190	150
					155	40,000						
						45,000						

CONFIGURATION: TWO 1000 LB BOMBS, TWO 110 GAL TANKS, TEN 5 IN. ROCKETS, OR ONE 1000 LB BOMB PLUS ONE 110 GAL TANK — GROSS WEIGHT: 12,200 TO 10,300 POUNDS							CONFIGURATION: TWO 1000 LB BOMBS, TWO 110 GAL TANKS TEN 5 IN. ROCKETS, OR ONE 1000 LB BOMB PLUS ONE 110 GAL TANK — GROSS WEIGHT: 10,300 POUNDS OR LESS					
APPROXIMATE				MP (IN. Hg) (2)	CAS (MPH)	PRESSURE ALTITUDE (FEET)	CAS (MPH)	MP (IN. Hg) (2)	APPROXIMATE			
RATE OF CLIMB	FROM SEA LEVEL DISTANCE	TIME	FUEL						FROM SEA LEVEL FUEL	TIME	DISTANCE	RATE OF CLIMB
1650	0	0	15(1)	61	175	SEA LEVEL	175	61	15(1)	0	0	2150
1550	10	3	23	61	185	5,000	185	61	21	2	8	2050
1450	22	6	32	61	190	10,000	190	61	28	5	16	1950
1150	36	10	43	F.T.	190	15,000	190	F.T.	36	8	27	1550
850	58	15	58	61	185	20,000	185	61	46	11	42	1250
750	84	22	75	61	180	25,000	180	61	58	15	60	1150
450	122	30	97	F.T.	175	30,000	175	F.T.	70	20	82	850
100	209	48	132	F.T.	165	35,000	165	F.T.	84	28	117	500
						40,000						
						45,000						

REMARKS:

1. Warm-up, taxi, and take-off allowance.
2. 3000 rpm.
3. Hold time at Military Power to a minimum.
4. Blower shift automatic.

LEGEND

RATE OF CLIMB - FEET PER MINUTE
DISTANCE - STATUTE MILES
TIME - MINUTES
FUEL - US. GALLONS
MP - MANIFOLD PRESSURE
CAS - CALIBRATED AIRSPEED
F.T. - FULL THROTTLE

DATA AS OF 8-1-53
BASED ON FLIGHT TEST

FUEL GRADE: 100/130
FUEL DENSITY: 6.0 LB/GAL

109-93-1775

Figure A-6. Military Power Climb (Sheet 1 of 2)

WADC
Form 241I
(11 Jun 51)

MILITARY POWER CLIMB CHART

STANDARD DAY

MODEL: F-51D ENGINE(S): (1) V-1650-7

CONFIGURATION: TWO 500 LB BOMBS OR TWO 75 GAL TANKS — GROSS WEIGHT: 11,200 TO 9800 POUNDS							CONFIGURATION: TWO 500 LB BOMBS OR TWO 75 GAL TANKS — GROSS WEIGHT: 9800 POUNDS OR LESS					
APPROXIMATE				MP (IN. Hg) (2)	CAS (MPH)	PRESSURE ALTITUDE (FEET)	CAS (MPH)	MP (IN. Hg) (2)	APPROXIMATE			
RATE OF CLIMB	FROM SEA LEVEL								FROM SEA LEVEL			RATE OF CLIMB
	DISTANCE	TIME	FUEL						FUEL	TIME	DISTANCE	
2000	0	0	15(1)	61	175	SEA LEVEL	175	61	15(1)	0	0	2400
1950	8	3	22	61	185	5,000	185	61	20	2	7	2450
1850	17	5	29	61	190	10,000	190	61	26	4	14	2300
1500	28	8	37	F.T.	190	15,000	190	F.T.	33	7	23	1900
1200	44	12	48	61	185	20,000	185	61	41	9	35	1600
1150	62	16	60	61	180	25,000	180	61	50	13	49	1500
800	83	21	73	F.T.	175	30,000	175	F.T.	60	16	66	1150
450	121	29	89	F.T.	165	35,000	165	F.T.	70	22	91	750
0	235	52	117	F.T.	155	40,000	155	F.T.	81	31	137	300
						45,000						

CONFIGURATION: WING RACKS — GROSS WEIGHT: 10,200 POUNDS OR LESS							CONFIGURATION: — GROSS WEIGHT:					
APPROXIMATE				MP (IN. Hg) (2)	CAS (MPH)	PRESSURE ALTITUDE (FEET)	CAS (MPH)	MP (IN. Hg)	APPROXIMATE			
RATE OF CLIMB	FROM SEA LEVEL								FROM SEA LEVEL			RATE OF CLIMB
	DISTANCE	TIME	FUEL						FUEL	TIME	DISTANCE	
2400	0	0	15(1)	61	175	SEA LEVEL						
2400	7	2	20	61	185	5,000						
2350	14	4	27	61	190	10,000						
2000	23	7	33	F.T.	190	15,000						
1700	35	9	41	61	185	20,000						
1600	48	12	50	61	180	25,000						
1250	64	16	58	F.T.	175	30,000						
850	86	61	68	F.T.	165	35,000						
450	124	68	77	F.T.	155	40,000						
						45,000						

REMARKS:

1. Warm-up, taxi, and take-off allowance.
2. 3000 rpm.
3. Hold time at Military Power to a minimum.
4. Blower shift automatic.

LEGEND

RATE OF CLIMB - FEET PER MINUTE
DISTANCE - STATUTE MILES
TIME - MINUTES
FUEL - US. GALLONS
MP - MANIFOLD PRESSURE
CAS - CALIBRATED AIRSPEED
F.T. - FULL THROTTLE

DATA AS OF 8-1-53
BASED ON FLIGHT TEST

FUEL GRADE: 100/130
FUEL DENSITY: 6.0 LB/GAL

109-93-1776

Figure A-6. Military Power Climb (Sheet 2 of 2)

WADC
Form 241Q
(11 Jun 51)

LANDING DISTANCES

(FEET)

STANDARD DAY

MODEL: F-51D ENGINE(S): (1) V-1650-7

GROSS WEIGHT (LB)	BEST IAS FOR APPROACH		HARD-SURFACE—NO WIND							
	POWER ON	POWER OFF	AT SEA LEVEL		AT 2000 FT		AT 4000 FT		AT 6000 FT	
	(MPH)	(MPH)	GROUND ROLL	TO CLEAR 50 FT OBST.	GROUND ROLL	TO CLEAR 50 FT OBST.	GROUND ROLL	TO CLEAR 50 FT OBST.	GROUND ROLL	TO CLEAR 50 FT OBST.
10,000	NOT AVAILABLE	140	1150	2000	1200	2050	1300	2150	1400	2250
9,000		140	1000	1850	1050	1900	1150	2000	1200	2050
8,000		140	850	1650	950	1750	1050	1850	1100	1900

REMARKS: 1. Landing distances are airplane requirements under normal service conditions.
2. Flaps full down.

LEGEND
IAS - INDICATED AIRSPEED
OBST. - OBSTACLE

DATA AS OF 8-1-53
BASED ON FLIGHT TEST

109-93-1779

FUEL GRADE: 100/130
FUEL DENSITY: 6.0 LB/GAL

Figure A-7. Landing Distances

COMBAT ALLOWANCE CHART

MILITARY POWER

STANDARD DAY

MODEL: F-51D ENGINE(S): (1) V-1650-7

PRESSURE ALTITUDE (FEET)	RPM	MP (IN. Hg)	BLOWER POSITION	MIXTURE POSITION	TIME LIMIT (MIN)	LIMIT COOLANT TEMP(°C)	FUEL FLOW (GPM)
SEA LEVEL	3000	61	LOW	NORMAL	15	121	2.5
2,000	3000	61	LOW	NORMAL	15	121	2.5
4,000	3000	61	LOW	NORMAL	15	121	3.0
6,000	3000	61	LOW	NORMAL	15	121	3.0
8,000	3000	61	LOW	NORMAL	15	121	3.0
10,000	3000	61	LOW	NORMAL	15	121	3.0
12,000	3000	61	LOW	NORMAL	15	121	3.0
14,000	3000	F.T.	LOW	NORMAL	15	121	3.0
16,000	3000	F.T.	LOW	NORMAL	15	121	2.5
18,000	3000	F.T.	LOW	NORMAL	15	121	2.5
20,000	3000	61	HIGH	NORMAL	15	121	3.0
22,000	3000	61	HIGH	NORMAL	15	121	3.0
24,000	3000	61	HIGH	NORMAL	15	121	3.0
26,000	3000	61	HIGH	NORMAL	15	121	3.0
28,000	3000	F.T.	HIGH	NORMAL	15	121	3.0
30,000	3000	F.T.	HIGH	NORMAL	15	121	2.5
32,000	3000	F.T.	HIGH	NORMAL	15	121	2.0
34,000	3000	F.T.	HIGH	NORMAL	15	121	2.0
36,000	3000	F.T.	HIGH	NORMAL	15	121	1.5
38,000	3000	F.T.	HIGH	NORMAL	15	121	1.5
40,000	3000	F.T.	HIGH	NORMAL	15	121	1.0

REMARKS:

1. F.T. = Full throttle.
2. Blower shift automatic.

DATA AS OF 8-20-44
BASED ON FLIGHT TEST

FUEL GRADE: 100/130
FUEL DENSITY: 6.0 LB/GAL

109-93-1771

Figure A-8. Combat Allowance

WADC
Form 241U
(11 Jun 51)

MAXIMUM ENDURANCE CHART

STANDARD DAY

MODEL: F-51D ENGINE(S): (1) V-1650-7

CONFIGURATION: SIX 5 IN. ROCKETS - PLUS TWO 75 GAL TANKS, TWO 110 GAL TANKS, TWO 1000 LB BOMBS, OR ONE 1000 LB BOMB PLUS ONE 110 GAL TANK GROSS WEIGHT: 13,000 TO 11,000 POUNDS							CONFIGURATION: SIX 5 IN. ROCKETS - PLUS TWO 75 GAL TANKS, TWO 110 GAL TANKS, TWO 1000 LB BOMBS, OR ONE 1000 LB BOMB PLUS ONE 110 GAL TANK GROSS WEIGHT: 11,000 POUNDS OR LESS			
APPROXIMATE							APPROXIMATE			
GPH	MIXTURE	RPM	MP (IN. Hg)	CAS (MPH)	PRESSURE ALTITUDE (FEET)	CAS (MPH)	MP (IN. Hg)	RPM	MIXTURE	GPH
48	NORMAL	1750	36	140	SEA LEVEL	140	30	1600	NORMAL	42
52	NORMAL	1750	35	140	5,000	140	30	1600	NORMAL	43
56	NORMAL	1800	36	140	10,000	140	31	1600	NORMAL	46
60	NORMAL	2050	35	140	15,000	140	30	1700	NORMAL	49
					20,000					
					25,000					
					30,000					
					35,000					
					40,000					
					45,000					

CONFIGURATION: TWO 1000 LB BOMBS, TWO 110 GAL TANKS, TEN 5 IN. ROCKETS, OR ONE 1000 LB BOMB PLUS ONE 110 GAL TANK GROSS WEIGHT: 12,200 TO 10,300 POUNDS							CONFIGURATION: TWO 1000 LB BOMBS, TWO 110 GAL TANKS, TEN 5 IN. ROCKETS, OR ONE 1000 LB BOMB PLUS ONE 110 GAL TANK GROSS WEIGHT: 10,300 POUNDS OR LESS			
APPROXIMATE							APPROXIMATE			
GPH	MIXTURE	RPM	MP (IN. Hg)	CAS (MPH)	PRESSURE ALTITUDE (FEET)	CAS (MPH)	MP (IN. Hg)	RPM	MIXTURE	GPH
43	NORMAL	1600	31	145	SEA LEVEL	135	29	1600	NORMAL	39
45	NORMAL	1600	31	145	5,000	135	29	1600	NORMAL	41
48	NORMAL	1600	32	145	10,000	135	29	1600	NORMAL	44
51	NORMAL	1750	F.T.	145	15,000	135	29	1700	NORMAL	47
56	NORMAL	2100	F.T.	145	20,000	135	F.T.	1950	NORMAL	51
					25,000					
					30,000					
					35,000					
					40,000					
					45,000					

REMARKS:

1. Use high blower for altitudes below heavy line.

LEGEND
GPH - FUEL CONSUMPTION
CAS - CALIBRATED AIRSPEED
F.T.- FULL THROTTLE

DATA AS OF: 8-1-53
BASED ON: FLIGHT TEST

FUEL GRADE: 100/130
FUEL DENSITY: 6.0 LB/GAL

109-93-1780

Figure A-9. Maximum Endurance (Sheet 1 of 2)

WADC
Form 241U
(11 Jun 51)

MAXIMUM ENDURANCE CHART

STANDARD DAY

MODEL: F-51D

ENGINE(S): (1) V-1650-7

CONFIGURATION (left): TWO 500 LB BOMBS OR TWO 75 GAL TANKS
GROSS WEIGHT: 11,200 TO 9800 POUNDS

CONFIGURATION (right): TWO 500 LB BOMBS OR TWO 75 GAL TANKS
GROSS WEIGHT: 9800 POUNDS OR LESS

APPROXIMATE				CAS (MPH)	PRESSURE ALTITUDE (FEET)	CAS (MPH)	APPROXIMATE			
GPH	MIXTURE	RPM	MP (IN. Hg)				MP (IN. Hg)	RPM	MIXTURE	GPH
40	NORMAL	1600	29	140	SEA LEVEL	140	26	1600	NORMAL	37
42	NORMAL	1600	29	140	5,000	140	26	1600	NORMAL	38
44	NORMAL	1600	29	140	10,000	140	26	1600	NORMAL	41
47	NORMAL	1700	29	140	15,000	140	26	1700	NORMAL	43
51	NORMAL	1950	F.T.	140	20,000	140	26	1800	NORMAL	47
					25,000	140	26	2000	NORMAL	50
					30,000					
					35,000					
					40,000					
					45,000					

CONFIGURATION (left): WING RACKS
GROSS WEIGHT: 10,200 POUNDS OR LESS

CONFIGURATION (right):
GROSS WEIGHT:

APPROXIMATE				CAS (MPH)	PRESSURE ALTITUDE (FEET)	CAS (MPH)	APPROXIMATE			
GPH	MIXTURE	RPM	MP (IN. Hg)				MP (IN. Hg)	RPM	MIXTURE	GPH
36	NORMAL	1600	25	140	SEA LEVEL					
37	NORMAL	1600	25	140	5,000					
39	NORMAL	1600	25	140	10,000					
41	NORMAL	1700	25	140	15,000					
44	NORMAL	1800	24	140	20,000					
46	NORMAL	2000	24	140	25,000					
51	NORMAL	2100	25	140	30,000					
					35,000					
					40,000					
					45,000					

REMARKS:

1. Use high blower for altitudes below heavy line.

LEGEND
GPH - FUEL CONSUMPTION
CAS - CALIBRATED AIRSPEED
F.T. - FULL THROTTLE

DATA AS OF: 8-1-53
BASED ON: FLIGHT TEST

FUEL GRADE: 100/130
FUEL DENSITY: 6.0 LB/GAL

109-93-1781

Figure A-9. Maximum Endurance (Sheet 2 of 2)

WADC Form 241K (11 Jun 51)	AIRCRAFT MODEL (S) F-51D ENGINE(S): (1) V-1650-7	FLIGHT OPERATION INSTRUCTION CHART STANDARD DAY CHART WEIGHT LIMITS: 10,200 TO 8000 POUNDS	EXTERNAL LOAD ITEMS WING RACKS ONLY

LIMITS	RPM	MP IN HG	BLOWER POSITION	MIXTURE POSITION	TIME LIMIT	COOLANT TEMP	TOTAL GPH
WAR EMER	3000	67	LOW	NORMAL	5 MIN	135°C	210
	3000	67	HIGH	NORMAL		135°C	210
MILITARY POWER	3000	61	LOW	NORMAL	15 MIN	135°C	180
	3000	61	HIGH	NORMAL		135°C	180

INSTRUCTIONS FOR USING CHART: Select figure in FUEL column equal to or less than amount of fuel to be used for cruising . Move horizontally to right or left and select RANGE value equal to or greater than the statute or nautical air miles to be flown. Vertically below and opposite value nearest desired cruising altitude (ALT), read rpm, manifold pressure (MP), and MIXTURE setting required. Refer to corresponding column and altitude for new power settings when gross weight falls below limits of this chart.

NOTES: Column I is for emergency high-speed cruising only. Columns II, III, IV, and V give progressive increase in range at a sacrifice in speed. Air miles per gallon (MI/GAL) (no wind), g llons per hr (GPH), and true airspeed (TAS) are approximate values for reference. Range values are for an average airplane flying alone (no wind)(1).

COLUMN I STATUTE	COLUMN I NAUTICAL	FUEL(1) US. GAL	COLUMN II STATUTE	COLUMN II NAUTICAL	COLUMN III STATUTE	COLUMN III NAUTICAL	COLUMN IV STATUTE	COLUMN IV NAUTICAL	FUEL(1) US. GAL	COLUMN V STATUTE	COLUMN V NAUTICAL
				SUBTRACT FUEL ALLOWANCES NOT AVAILABLE FOR CRUISING.							
880	765	240	985	855	1105	960	1200	1040	240	1330	1155
805	700	220	900	780	1010	875	1100	955	220	1220	1060
730	635	200	820	710	920	800	1000	870	200	1110	965
640	555	180	740	645	830	720	900	780	180	1000	870
585	510	160	655	570	735	640	800	695	160	890	775
515	445	140	575	500	645	560	700	610	140	775	675
440	380	120	490	425	550	475	600	520	120	665	580
365	315	100	410	355	460	400	500	435	100	555	480
295	255	80	330	285	370	320	400	350	80	445	385
220	190	60	245	210	275	240	300	260	60	335	290
145	125	40	165	145	185	160	200	175	40	220	190
75	65	20	80	70	90	78	100	87	20	110	95

MAXIMUM CONTINUOUS RPM	MP IN.	MIX-TURE	TOTAL GPH	TAS MPH	TAS KN	PRESS. ALT FEET	(4.10 STAT. (3.56 NAUT.) MI/GAL) RPM	MP IN.	MIX-TURE	TOTAL GPH	TAS MPH	TAS KN	(4.60 STAT. (4.00 NAUT.) MI/GAL) RPM	MP IN.	MIX-TURE	TOTAL GPH	TAS MPH	TAS KN	(5.00 STAT. (4.34 NAUT.) MI/GAL) RPM	MP IN.	MIX-TURE	TOTAL GPH	TAS MPH	TAS KN	PRESS. ALT FEET	MAXIMUM AIR RANGE RPM	MP IN.	MIX-TURE	TOTAL GPH	TAS MPH	TAS KN
						40,000																			40,000						
						35,000													2700	F.T.	NORMAL	80	408	354	35,000	2500	F.T	NORMAL	66	367	319
	SEE COLUMN II					30,000	2700	F.T.	NORMAL	97	424	368	2650	F.T.	NORMAL	90	417	362	2450	F.T.	NORMAL	77	391	340	30,000	2250	F.T	NORMAL	62	348	302
	SEE COLUMN II					25,000	2700	46.0	NORMAL	98	413	358	2400	42.0	NORMAL	86	397	345	2250	F.T.	NORMAL	73	370	321	25,000	1850	F.T	NORMAL	59	335	291
	SEE COLUMN II					20,000	2700	46.0	NORMAL	93	389	338	2400	42.5	NORMAL	80	371	322	2100	36.5	NORMAL	67	346	300	20,000	2100	F.T	NORMAL	57	316	274
2700	46	NORMAL	103	384	334	15,000	2500	F.T.	NORMAL	90	373	324	2300	F.T.	NORMAL	76	351	305	2100	F.T.	NORMAL	65	326	283	15,000	1800	F.T	NORMAL	52	290	252
2700	46	NORMAL	98	361	314	10,000	2500	42.5	NORMAL	83	346	300	2200	40.0	NORMAL	71	325	282	1900	37.0	NORMAL	60	300	261	10,000	1600	31.0	NORMAL	47	261	227
2700	46	NORMAL	91	339	286	5,000	2500	43.0	NORMAL	79	323	280	2200	40.0	NORMAL	66	303	263	1850	36.5	NORMAL	55	278	242	5,000	1600	28.5	NORMAL	42	232	202
2700	46	NORMAL	86	315	274	SL	2500	43.0	NORMAL	74	300	261	2150	39.5	NORMAL	61	279	242	1800	36.0	NORMAL	50	255	222	SL		SEE COLUMN IV				

SPECIAL NOTES

(1) Make allowance for warm-up, take-off, and climb plus allowance for wind, reserve, and combat as required.
(2) High blower above heavy line.

EXAMPLE

At 9500 lb gross weight with 80 gal of fuel (after deducting total allowances of 20 gal) to fly 330 stat. air miles at 5000 ft altitude maintain 2500 rpm and 43 in. manifold pressure with mixture set: NORMAL.

LEGEND

ALT - PRESSURE ALTITUDE
MP - MANIFOLD PRESSURE
GPH - US. GALLONS PER HOUR
TAS - TRUE AIRSPEED
KN - KNOTS
SL - SEA LEVEL
F.T. - FULL THROTTLE

REVISED 1-22-47
DATA AS OF 8-20-44
BASED ON: FLIGHT TESTS

F-51D-1-93-1

FUEL GRADE: 100/130 FUEL DENSITY: 6.0 LB/GAL

Figure A-10. Flight Operation Instruction Chart—No External Load—10,200 to 8000 Pounds

WADC Form 241K (11 Jun 51)

FLIGHT OPERATION INSTRUCTION CHART

STANDARD DAY

AIRCRAFT MODEL (S): F-51D

ENGINE(S): (1) V-1650-7

CHART WEIGHT LIMITS: 11,200 TO 9800 POUNDS

EXTERNAL LOAD ITEMS

TWO 500 LB BOMBS OR
TWO 75 GAL DROP TANKS

LIMITS	RPM	MP IN. HG	BLOWER POSITION	MIXTURE POSITION	TIME LIMIT	COOLANT TEMP	TOTAL GPH
WAR EMER	3000	67	LOW	NORMAL	5 MIN	135°C	210
	3000	67	HIGH	NORMAL		135°C	210
MILITARY POWER	3000	61	LOW	NORMAL	15 MIN	135°C	180
	3000	61	HIGH	NORMAL		135°C	180

INSTRUCTIONS FOR USING CHART: Select figure in FUEL column equal to or less than amount of fuel to be used for cruising(1). Move horizontally to right or left and select RANGE value equal to or greater than the statute or nautical air miles to be flown. Vertically below and opposite value nearest desired cruising altitude (ALT), read rpm, manifold pressure (MP), and MIXTURE setting required. Refer to corresponding column and altitude for new power settings when gross weight falls below limits of this chart.

NOTES: Column I is for emergency high-speed cruising only. Columns II, III, IV, and V give progressive increase in range at a sacrifice in speed. Air miles per gallon (MI/GAL) (no wind), gallons per hr (GPH), and true airspeed (TAS) are approximate values for reference. Range values are for an average airplane flying alone (no wind)(1).

COLUMN I		FUEL(1)	COLUMN II		COLUMN III		COLUMN IV		FUEL(1)	COLUMN V	
RANGE IN AIR MILES		US. GAL	RANGE IN AIR MILES		RANGE IN AIR MILES		RANGE IN AIR MILES		US. GAL	RANGE IN AIR MILES	
STATUTE	NAUTICAL		STATUTE	NAUTICAL	STATUTE	NAUTICAL	STATUTE	NAUTICAL		STATUTE	NAUTICAL
			SUBTRACT FUEL ALLOWANCES NOT AVAILABLE FOR CRUISING(1)								
1320	1145	390	1430	1240	1600	1390	1765	1535	390	1895	1650
1250	1085	370	1350	1175	1515	1315	1670	1450	370	1795	1560
1185	1030	350	1280	1110	1435	1250	1580	1370	350	1700	1480
1120	975	330	1200	1040	1350	1170	1490	1295	330	1600	1390
1050	910	310	1130	980	1265	1100	1400	1220	310	1500	1305
985	855	290	1050	915	1180	1050	1305	1135	290	1400	1220
915	795	270	980	850	1100	955	1215	1055	270	1300	1130
845	735	250	900	785	1015	885	1115	970	250	1200	1045
780	675	230	830	720	930	810	1030	895	230	1110	965
710	615	210	755	655	850	740	945	820	210	1010	880
645	560	190	685	595	770	670	855	745	190	915	795
575	500	170	610	530	690	600	765	665	170	820	715
510	445	150	540	470	605	525	675	585	150	720	625

MAXIMUM CONTINUOUS						PRESS. ALT	(3.50 STAT. (3.13 NAUT.) MI/GAL)						(4.05 STAT. (3.52 NAUT.) MI/GAL)						(4.49 STAT. (3.90 NAUT.) MI/GAL)						PRESS. ALT	MAXIMUM AIR RANGE					
RPM	MP IN.	MIX-TURE	TOTAL GPH	APPROX TAS MPH	APPROX TAS KN	FEET	RPM	MP IN.	MIX-TURE	TOTAL GPH	APPROX TAS MPH	APPROX TAS KN	RPM	MP IN.	MIX-TURE	TOTAL GPH	APPROX TAS MPH	APPROX TAS KN	RPM	MP IN.	MIX-TURE	TOTAL GPH	APPROX TAS MPH	APPROX TAS KN	FEET	RPM	MP IN.	MIX-TURE	TOTAL GPH	APPROX TAS MPH	APPROX TAS KN
						40,000																			40,000						
						35,000																			35,000						
	SEE COLUMN III					30,000							2700	F. T.	NORMAL	94	383	333	2500	F. T.	NORMAL	79	356	310	30,000	2300	F. T.	NORMAL	63	315	274
	SEE COLUMN II					25,000	2700	46.0	NORMAL	98	377	328	2550	44.0	NORMAL	90	367	319	2300	F. T.	NORMAL	76	341	297	25,000	2100	F. T.	NORMAL	60	305	265
	SEE COLUMN II					20,000	2700	46.0	NORMAL	93	355	309	2500	44.0	NORMAL	85	344	299	2150	38.0	NORMAL	70	319	277	20,000	2200	F. T.	NORMAL	61	294	256
2700	46	NORMAL	103	356	310	15,000	2600	44.5	NORMAL	97	349	303	2400	F. T.	NORMAL	80	327	284	2200	F. T.	NORMAL	68	303	263	15,000	1950	F. T.	NORMAL	57	277	241
2700	46	NORMAL	98	334	290	10,000	2600	44.5	NORMAL	91	326	284	2300	41.0	NORMAL	75	304	264	2000	38.0	NORMAL	63	281	244	10,000	1650	34.5	NORMAL	52	252	219
2700	46	NORMAL	91	312	271	5,000	2600	44.0	NORMAL	84	304	264	2300	41.0	NORMAL	70	283	246	1950	37.5	NORMAL	58	260	226	5,000	1600	34.0	NORMAL	48	230	200
2700	46	NORMAL	86	292	254	SL	2550	44.0	NORMAL	78	282	245	2250	40.5	NORMAL	65	262	228	1900	37.0	NORMAL	53	239	208	SL	1600	31.0	NORMAL	42	202	176

SPECIAL NOTES

(1) Make allowance for warm-up, take-off, and climb plus allowance for wind, reserve, and combat as required.
(2) High blower above heavy line.

EXAMPLE

At 10,000 lb gross weight with 210 gal of fuel (after deducting total allowances of 20 gal) to fly 850 stat. air miles at 5000 ft altitude, maintain 2300 rpm and 41 in. manifold pressure with mixture set: NORMAL.

LEGEND

ALT - PRESSURE ALTITUDE
MP - MANIFOLD PRESSURE
GPH – US. GALLONS PER HOUR
TAS - TRUE AIRSPEED
KN - KNOTS
SL - SEA LEVEL
F.T. - FULL THROTTLE

REVISED 1-21-47
DATA AS OF: 9-10-44
BASED ON: FLIGHT TEST

F-51D-1-93-2

FUEL GRADE: 100/130

FUEL DENSITY: 6.0 LB/GAL

Figure A-11. Flight Operation Instruction Chart—With External Load—11,200 to 9800 Pounds

WADC Form 241K (11 Jun 51)

AIRCRAFT MODEL (S)

F-51D

ENGINE(S): (1) V-1650-7

FLIGHT OPERATION INSTRUCTION CHART

STANDARD DAY

CHART WEIGHT LIMITS: 9800 TO 8100 POUNDS

EXTERNAL LOAD ITEMS

TWO 500 LB BOMBS OR
TWO 75 GAL DROP TANKS

LIMITS	RPM	MP IN. HG	BLOWER POSITION	MIXTURE POSITION	TIME LIMIT	COOLANT TEMP	TOTAL GPH
WAR EMER	3000	67	LOW	NORMAL	5 MIN	135°C	210
	3000	67	HIGH	NORMAL		135°C	210
MILITARY POWER	3000	61	LOW	NORMAL	15 MIN	135°C	180
	3000	61	HIGH	NORMAL		135°C	180

INSTRUCTIONS FOR USING CHART: Select figure in FUEL column equal to or less than amount of fuel to be used for cruising(1). Move horizontally to right or left and select RANGE value equal to or greater than the statute or nautical air miles to be flown. Vertically below and opposite value nearest desired cruising altitude (ALT), read rpm, manifold pressure (MP), and MIXTURE setting required. Refer to corresponding column and altitude for new power settings when gross weight falls below limits of this chart.

NOTES: Column I is for emergency high-speed cruising only. Columns II, III, IV, and V give progressive increase in range at a sacrifice in speed. Air miles per gallon (MI/GAL) (no wind), gallons per hr (GPH), and true airspeed (TAS) are approximate values for reference. Range values are for an average airplane flying alone (no wind)(1).

COLUMN I		FUEL(1)	COLUMN II		COLUMN III		COLUMN IV		FUEL(1)	COLUMN V	
RANGE IN AIR MILES		US. GAL	RANGE IN AIR MILES		RANGE IN AIR MILES		RANGE IN AIR MILES		US. GAL	RANGE IN AIR MILES	
STATUTE	NAUTICAL		STATUTE	NAUTICAL	STATUTE	NAUTICAL	STATUTE	NAUTICAL		STATUTE	NAUTICAL
			SUBTRACT FUEL ALLOWANCES NOT AVAILABLE FOR CRUISING(1)								
950	825	280	1050	915	1170	1020	1275	1110	280	1380	1200
880	765	260	975	845	1090	945	1185	1030	260	1280	1110
815	710	240	900	785	1005	875	1090	945	240	1185	1030
745	645	220	825	715	920	800	1000	870	220	1085	940
680	590	200	750	650	835	725	910	790	200	990	860
610	530	180	675	585	755	655	820	715	180	890	775
545	475	160	600	520	670	580	730	635	160	790	685
475	415	140	525	455	585	510	635	550	140	690	600
405	350	120	450	390	500	435	545	475	120	590	515
340	295	100	375	325	420	365	455	395	100	495	430
270	235	80	300	260	335	290	365	315	80	395	345
205	180	60	225	195	250	220	275	240	60	295	255
135	115	40	150	130	165	145	180	155	40	195	170

MAXIMUM CONTINUOUS						PRESS. ALT	(3.75 STAT. (3.25 NAUT.) MI/GAL)						(4.18 STAT. (3.64 NAUT.) MI/GAL)						(4.55 STAT. (3.96 NAUT.) MI/GAL)						PRESS. ALT	MAXIMUM AIR RANGE					
RPM	MP IN.	MIX-TURE	TOTAL GPH	APPROX TAS MPH	APPROX TAS KN	FEET	RPM	MP IN.	MIX-TURE	TOTAL GPH	APPROX TAS MPH	APPROX TAS KN	RPM	MP IN.	MIX-TURE	TOTAL GPH	APPROX TAS MPH	APPROX TAS KN	RPM	MP IN.	MIX-TURE	TOTAL GPH	APPROX TAS MPH	APPROX TAS KN	FEET	RPM	MP IN.	MIX-TURE	TOTAL GPH	APPROX TAS MPH	APPROX TAS KN
						40,000																			40,000						
SEE COLUMN IV						35,000													2700	F.T.	NORMAL	80	370	322	35,000	2450	F.T.	NORMAL	65	327	284
SEE COLUMN III						30,000							2650	F.T.	NORMAL	90	381	331	2500	F.T.	NORMAL	78	360	313	30,000	2300	F.T.	NORMAL	65	326	283
SEE COLUMN II						25,000	2700	46.0	NORMAL	98	377	328	2450	42.5	NORMAL	87	362	314	2300	F.T.	NORMAL	75	342	297	25,000	2100	F.T.	NORMAL	62	311	270
SEE COLUMN II						20,000	2700	46.0	NORMAL	93	355	308	2400	42.5	NORMAL	80	337	293	2150	38	NORMAL	70	318	276	20,000	2150	F.T.	NORMAL	59	294	256
2700	46	NORMAL	103	356	309	15,000	2550	43.0	NORMAL	91	341	296	2350	F.T.	NORMAL	79	322	280	2150	F.T.	NORMAL	66	300	261	15,000	1900	F.T.	NORMAL	55	272	236
2700	46	NORMAL	98	333	289	10,000	2500	43.0	NORMAL	83	316	275	2200	40.0	NORMAL	71	298	259	1950	37	NORMAL	61	278	242	10,000	1600	34.0	NORMAL	50	251	218
2700	46	NORMAL	91	311	270	5,000	2450	42.5	NORMAL	77	293	255	2200	40.0	NORMAL	66	276	240	1900	37	NORMAL	55	256	222	5,000	1600	32.5	NORMAL	46	229	199
2700	46	NORMAL	86	291	253	SL	2450	42.5	NORMAL	72	273	237	2150	39.5	NORMAL	61	256	222	1850	37		51	236	205	SL	1600	29.5	NORMAL	41	200	174

SPECIAL NOTES

(1) Make allowance for warm-up, take-off, and climb plus allowance for wind, reserve, and combat as required.
(2) High blower above heavy line.

EXAMPLE

At 9400 lb gross weight with 100 gal of fuel (after deducting total allowances of 20 gal) to fly 420 stat. air miles at 9000 ft altitude, maintain 2200 rpm and 40 in. manifold pressure with mixture set: NORMAL.

LEGEND

ALT - PRESSURE ALTITUDE
MP - MANIFOLD PRESSURE
GPH - US. GALLONS PER HOUR
TAS - TRUE AIRSPEED
KN - KNOTS
SL - SEA LEVEL
F.T. - FULL THROTTLE

REVISED 1-21-47
DATA AS OF: 9-10-44
BASED ON: FLIGHT TEST

F-51D-1-93-3

FUEL GRADE: 100/130

FUEL DENSITY: 6.0 LB/GAL

Figure A-12. Flight Operation Instruction Chart—With External Load—9800 to 8100 Pounds

WADC Form 241K (11 Jun 51)

FLIGHT OPERATION INSTRUCTION CHART

STANDARD DAY

AIRCRAFT MODEL (S): F-51D

ENGINE(S): (1) V-1650-7

CHART WEIGHT LIMITS: 13,000 TO 11,000 POUNDS

EXTERNAL LOAD ITEMS

SIX ROCKETS PLUS TWO 75 GAL DROP TANKS,
SIX ROCKETS PLUS TWO 110 GAL DROP TANKS,
SIX ROCKETS PLUS ONE 110 GAL DROP TANK AND ONE 1000 LB BOMB, OR
SIX ROCKETS PLUS TWO 1000 LB BOMBS

LIMITS	RPM	MP IN. HG	BLOWER POSITION	MIXTURE POSITION	TIME LIMIT	COOLANT TEMP	TOTAL GPH
WAR EMER	3000	67	LOW	NORMAL	5 MIN	135°C	210
	3000	67	HIGH	NORMAL		135°C	210
MILITARY POWER	3000	61	LOW	NORMAL	15 MIN	135°C	180
	3000	61	HIGH	NORMAL		135°C	180

INSTRUCTIONS FOR USING CHART: Select figure in FUEL column equal to or less than amount of fuel to be used for cruising(1). Move horizontally to right or left and select RANGE value equal to or greater than the statute or nautical air miles to be flown. Vertically below and opposite value nearest desired cruising altitude (ALT), read rpm, manifold pressure (MP), and MIXTURE setting required. Refer to corresponding column and altitude for new power settings when gross weight falls below limits of this chart.

NOTES: Column I is for emergency high-speed cruising only. Columns II, III, IV, and V give progressive increase in range at a sacrifice in speed. Air miles per gallon (MI/GAL) (no wind), gallons per hr (GPH), and true airspeed (TAS) are approximate values for reference. Range values are for an average airplane flying alone (no wind)(1).

COLUMN I Range in air miles Statute	COLUMN I Nautical	FUEL(1) US. GAL	COLUMN II Statute	COLUMN II Nautical	COLUMN III Statute	COLUMN III Nautical	COLUMN IV Statute	COLUMN IV Nautical	FUEL(1) US. GAL	COLUMN V Statute	COLUMN V Nautical
			SUBTRACT FUEL ALLOWANCES NOT AVAILABLE FOR CRUISING.(1)								
1395	1210	450	1485	1290	1625	1410	1730	1505	450	1810	1570
1300	1130	420	1380	1200	1510	1310	1600	1390	420	1680	1460
1205	1045	390	1280	1110	1390	1210	1475	1280	390	1550	1350
1110	965	360	1175	1020	1275	1110	1345	1170	360	1415	1230
1015	880	330	1070	930	1160	1010	1220	1060	330	1290	1120
925	805	300	970	845	1055	915	1110	965	300	1170	1015
830	720	270	875	760	950	825	1000	870	270	1055	915
740	645	240	780	680	840	730	890	775	240	935	810
645	560	210	680	590	740	645	775	675	210	820	715
555	480	180	585	510	630	550	665	580	180	700	610
460	400	150	485	420	525	455	555	480	150	585	510
370	320	120	390	340	420	365	445	385	120	470	410

MAXIMUM CONTINUOUS RPM	MP IN.	MIX-TURE	TOTAL GPH	TAS MPH	TAS KN	PRESS. ALT FEET	(3.24 STAT. (2.82 NAUT.) MI/GAL) RPM	MP IN.	MIX-TURE	TOTAL GPH	TAS MPH	TAS KN	(3.51 STAT. (3.05 NAUT.) MI/GAL) RPM	MP IN.	MIX-TURE	TOTAL GPH	TAS MPH	TAS KN	(3.70 STAT. (3.22 NAUT.) MI/GAL) RPM	MP IN.	MIX-TURE	TOTAL GPH	TAS MPH	TAS KN	PRESS. ALT FEET	MAXIMUM AIR RANGE RPM	MP IN.	MIX-TURE	TOTAL GPH	TAS MPH	TAS KN
						40,000																			40,000						
						35,000																			35,000						
						30,000																			30,000						
SEE COLUMN II						25,000	2700	46.0	NORMAL	98	332	288	2550	44.0	NORMAL	92	323	281	2350	F.T.	NORMAL	80	295	256	25,000						
SEE COLUMN II						20,000	2700	46.0	NORMAL	93	314	273	2550	44.5	NORMAL	86	304	264	2350	41	NORMAL	77	285	248	20,000						
2700	46	NORMAL	103	322	280	15,000	2600	44.5	NORMAL	97	314	273	2450	F.T.	NORMAL	85	299	260	2350	F.T.	NORMAL	77	284	247	15,000	2100	F.T.	NORMAL	63	245	213
2700	46	NORMAL	98	302	262	10,000	2600	44.5	NORMAL	91	295	256	2400	42.0	NORMAL	80	281	244	2250	40	NORMAL	72	267	232	10,000	1950	37.5	NORMAL	61	240	209
2700	46	NORMAL	91	283	246	5,000	2600	44.5	NORMAL	85	275	239	2400	42.0	NORMAL	74	261	227	2200	40	NORMAL	66	247	215	5,000	1950	36.0	NORMAL	57	223	194
2700	46	NORMAL	86	265	230	SL	2600	44.0	NORMAL	78	256	222	2350	41.5	NORMAL	69	243	211	2200	40	NORMAL	62	231	201	SL	1900	37.0	NORMAL	53	205	178

SPECIAL NOTES

(1) Make allowance for warm-up, take-off, and climb plus allowance for wind, reserve, and combat as required.
(2) High blower above heavy line.

EXAMPLE

At 11,800 lb gross weight with 330 gal of fuel (after deducting total allowances of 26 gal) to fly 1220 stat. air miles at 10,000 ft altitude, maintain 2250 rpm and 40 in. manifold pressure with mixture set: NORMAL.

LEGEND

ALT - PRESSURE ALTITUDE
MP - MANIFOLD PRESSURE
GPH - US. GALLONS PER HOUR
TAS - TRUE AIRSPEED
KN - KNOTS
SL - SEA LEVEL
F.T. - FULL THROTTLE

REVISED 1-23-47
DATA AS OF: 12-1-44
BASED ON: FLIGHT TEST

F-51D-1-93-4

FUEL GRADE: 100/130

FUEL DENSITY: 6.0 LB/GAL

Figure A-13. Flight Operation Instruction Chart—With External Load—13,000 to 11,000 Pounds

WADC Form 241K (11 Jun 51)

AIRCRAFT MODEL (S)

F-51D

ENGINE(S): (1) V-1650-7

FLIGHT OPERATION INSTRUCTION CHART

STANDARD DAY

CHART WEIGHT LIMITS: 11,000 TO 8900 POUNDS

EXTERNAL LOAD ITEMS

SIX ROCKETS PLUS TWO 75 GAL DROP TANKS,
SIX ROCKETS PLUS TWO 110 GAL DROP TANKS,
SIX ROCKETS PLUS ONE 110 GAL DROP TANK AND ONE 1000 LB BOMB, OR
SIX ROCKETS PLUS TWO 1000 LB BOMBS

LIMITS	RPM	MP IN HG	BLOWER POSITION	MIXTURE POSITION	TIME LIMIT	COOLANT TEMP	TOTAL GPH
WAR EMER	3000 3000	67 67	LOW HIGH	NORMAL NORMAL	5 MIN	135°C 135°C	210 210
MILITARY POWER	3000 3000	61 61	LOW HIGH	NORMAL NORMAL	15 MIN	135°C 135°C	180 180

INSTRUCTIONS FOR USING CHART: Select figure in FUEL column equal to or less than amount of fuel to be used for cruising (1). Move horizontally to right or left and select RANGE value equal to or greater than the statute or nautical air miles to be flown. Vertically below and opposite value nearest desired cruising altitude (ALT), read rpm, manifold pressure (MP), and MIXTURE setting required. Refer to corresponding column and altitude for new power settings when gross weight falls below limits of this chart.

NOTES: Column I is for emergency high-speed cruising only. Columns II, III, IV, and V give progressive increase in range at a sacrifice in speed. Air miles per gallon (MI/GAL) (no wind), gallons per hr (GPH), and true airspeed (TAS) are approximate values for reference. Range values are for an average airplane flying alone (no wind)(1).

COLUMN I — RANGE IN AIR MILES — STATUTE	COLUMN I — NAUTICAL	FUEL(1) US. GAL	COLUMN II — STATUTE	COLUMN II — NAUTICAL	COLUMN III — STATUTE	COLUMN III — NAUTICAL	COLUMN IV — STATUTE	COLUMN IV — NAUTICAL	FUEL(1) US. GAL	COLUMN V — STATUTE	COLUMN V — NAUTICAL
			SUBTRACT FUEL ALLOWANCES NOT AVAILABLE FOR CRUISING(1)								
1040 950	905 825	330 300	1170 1065	1015 925	1285 1170	1115 1020	1400 1275	1220 1110	330 300	1440 1310	1250 1140
855 760 665	745 660 580	270 240 210	960 850 745	835 740 645	1050 935 820	910 815 715	1150 1020 895	1000 885 775	270 240 210	1180 1050 915	1025 915 795
520 475 380	450 415 330	180 150 120	640 530 425	555 460 370	700 585 470	610 510 410	765 635 510	665 550 445	180 150 120	785 655 525	680 570 455
285 190 95	250 165 83	90 60 30	320 215 105	280 185 91	350 235 115	305 205 100	380 255 125	330 220 110	90 60 30	390 260 130	425 226 115

MAXIMUM CONTINUOUS; COLUMN II (3.55 STAT. (3.08 NAUT.) MI/GAL):

PRESS. ALT FEET	MAX CONT RPM	MP IN.	MIXTURE	APPROX TOTAL LB HR	TAS MPH	TAS KN	COL II RPM	MP IN.	MIXTURE	APPROX TOTAL LB HR	TAS MPH	TAS KN
40,000												
35,000												
30,000												
25,000	SEE COLUMN II						2700	46.0	NORMAL	98	350	304
20,000	SEE COLUMN II						2650	45.5	NORMAL	91	327	284
15,000	2700	46	NORMAL	103	332	288	2500	F.T.	NORMAL	89	316	275
10,000	2700	46	NORMAL	98	311	270	2500	43.0	NORMAL	83	295	256
5,000	2700	46	NORMAL	91	290	252	2450	42.5	NORMAL	77	273	237
SL	2700	46	NORMAL	86	272	236	2400	42.0	NORMAL	71	254	221

COLUMN III (3.90 STAT. (3.39 NAUT.) MI/GAL); COLUMN IV (4.25 STAT. (3.70 NAUT.) MI/GAL); COLUMN V MAXIMUM AIR RANGE:

PRESS. ALT FEET	COL III RPM	MP IN.	MIXTURE	APPROX TOTAL LB HR	TAS MPH	TAS KN	COL IV RPM	MP IN.	MIXTURE	APPROX TOTAL LB HR	TAS MPH	TAS KN	MAX AIR RANGE RPM	MP IN.	MIXTURE	APPROX TOTAL LB HR	TAS MPH	TAS KN
40,000																		
35,000																		
30,000																		
25,000	2350	F.T.	NORMAL	81	322	280	2150	F.T.	NORMAL	64	276	240						
20,000	2300	41.0	NORMAL	77	303	263	2000	34.5	NORMAL	63	273	237	2150	F.T.	NORMAL	60	261	227
15,000	2300	F.T.	NORMAL	74	292	254	2100	F.T.	NORMAL	63	269	234	1950	F.T.	NORMAL	57	253	220
10,000	2150	39.5	NORMAL	69	273	237	1900	37.0	NORMAL	60	253	220	1650	34.5	NORMAL	52	225	196
5,000	2100	39.0	NORMAL	65	252	219	1850	36.5	NORMAL	55	233	202	1600	34.0	NORMAL	48	210	183
SL	2050	38.5	NORMAL	57	233	202	1800	36.0	NORMAL	50	212	184	SEE COLUMN IV					

SPECIAL NOTES

(1) Make allowance for warm-up, take-off, and climb plus allowance for wind, reserve, and combat as required.
(2) High blower above heavy line.

EXAMPLE

At 10,600 lb gross weight with 270 gal of fuel (after deducting total allowances of 26 gal) to fly 1150 stat. air miles at 10,000 ft altitude, maintain 1900 rpm and 37 in. manifold pressure with mixture set: NORMAL.

LEGEND

ALT - PRESSURE ALTITUDE
MP - MANIFOLD PRESSURE
GPH - US. GALLONS PER HOUR
TAS - TRUE AIRSPEED
KN - KNOTS
SL - SEA LEVEL
F.T. - FULL THROTTLE

REVISED 1-23-47
DATA AS OF: 12-1-44
BASED ON: FLIGHT TEST

F-51D-1-93-5

FUEL GRADE: 100/130

FUEL DENSITY: 6.0 LB/GAL

Figure A-14. Flight Operation Instruction Chart—With External Load—11,000 to 8900 Pounds

WADC Form 241K, 11 Jun 51

FLIGHT OPERATION INSTRUCTION CHART

STANDARD DAY

AIRCRAFT MODEL (S): F-51D

ENGINE(S): (1) V-1650-7

CHART WEIGHT LIMITS: 12,200 TO 10,300 POUNDS

EXTERNAL LOAD ITEMS

TWO 1000 LB BOMBS,
ONE 1000 LB BOMB PLUS ONE 110 GAL DROP TANK,
TWO 110 GAL DROP TANKS, OR TEN 5 IN. ROCKETS

LIMITS	RPM	MP IN. HG.	BLOWER POSITION	MIXTURE POSITION	TIME LIMIT	COOLANT TEMP	TOTAL GPH
WAR EMER	3000 3000	67 67	LOW HIGH	NORMAL NORMAL	5 MIN	135°C 135°C	210 210
MILITARY POWER	3000 3000	61 61	LOW HIGH	NORMAL NORMAL	15 MIN	135°C 135°C	180 180

INSTRUCTIONS FOR USING CHART: Select figure in FUEL column equal to or less than amount of fuel to be used for cruising (1). Move horizontally to right or left and select RANGE value equal to or greater than the statute or nautical air miles to be flown. Vertically below and opposite value nearest desired cruising altitude (ALT), read rpm, manifold pressure (MP), and MIXTURE setting required. Refer to corresponding column and altitude for new power settings when gross weight falls below limits of this chart.

NOTES: Column I is for emergency high-speed cruising only. Columns II, III, IV, and V give progressive increase in range at a sacrifice in speed. Air miles per gallon (MI/GAL) (no wind), gallons per hr (GPH), and true airspeed (TAS) are approximate values for reference. Range values are for an average airplane flying alone (no wind) (1).

COLUMN I		FUEL (1) US. GAL	COLUMN II		COLUMN III		COLUMN IV		FUEL (1) US. GAL	COLUMN V	
RANGE IN AIR MILES			RANGE IN AIR MILES		RANGE IN AIR MILES		RANGE IN AIR MILES			RANGE IN AIR MILES	
STATUTE	NAUTICAL		STATUTE	NAUTICAL	STATUTE	NAUTICAL	STATUTE	NAUTICAL		STATUTE	NAUTICAL
			SUBTRACT FUEL ALLOWANCES NOT AVAILABLE FOR CRUISING (1)								
1455	1265	440	1595	1385	1785	1550	1960	1700	440	2055	1780
1325	1150	400	1445	1255	1615	1405	1770	1540	400	1855	1610
1190	1035	360	1295	1125	1445	1255	1580	1370	360	1660	1440
1060	920	320	1145	995	1275	1110	1390	1210	320	1460	1270
925	805	280	1000	870	1115	970	1220	1060	280	1280	1110
795	690	240	855	745	955	830	1045	910	240	1095	950
660	575	200	715	620	795	690	870	755	200	915	795
530	460	160	570	495	635	550	695	605	160	730	635

MAXIMUM CONTINUOUS						PRESS ALT FEET	(3.57 STAT. (3.10 NAUT.) MI/GAL)						(3.98 STAT. (3.46 NAUT.) MI/GAL)						(4.35 STAT. (3.78 NAUT.) MI/GAL)						PRESS ALT FEET	MAXIMUM AIR RANGE					
RPM	MP IN.	MIX-TURE	APPROX TOTAL GPH	TAS MPH	TAS KN		RPM	MP IN.	MIX-TURE	APPROX TOTAL GPH	TAS MPH	TAS KN	RPM	MP IN.	MIX-TURE	APPROX TOTAL GPH	TAS MPH	TAS KN	RPM	MP IN.	MIX-TURE	APPROX TOTAL GPH	TAS MPH	TAS KN		RPM	MP IN.	MIX-TURE	APPROX TOTAL GPH	TAS MPH	TAS KN
						40,000																			40,000						
						35,000																			35,000						
						30,000																			30,000						
	SEE COLUMN II					25,000	2700	46	NORMAL	98	370	322	2500	43.5	NORMAL	89	356	309	2250	F.T.	NORMAL	72	313	272	25,000						
	SEE COLUMN II					20,000	2700	46	NORMAL	93	349	303	2500	43.5	NORMAL	84	335	291	2150	37.5	NORMAL	70	304	264	20,000	2200	F.T.	NORMAL	61	280	243
2700	46	NORMAL	103	352	306	15,000	2600	44	NORMAL	95	343	298	2400	F.T.	NORMAL	81	322	280	2150	F.T.	NORMAL	66	291	253	15,000	2000	F.T.	NORMAL	59	271	236
2700	46	NORMAL	98	327	284	10,000	2550	44	NORMAL	89	317	276	2250	40.5	NORMAL	74	297	258	1950	37.5	NORMAL	62	272	236	10,000	1800	36.0	NORMAL	56	253	220
2700	46	NORMAL	91	308	268	5,000	2550	44	NORMAL	82	298	259	2250	40.5	NORMAL	69	278	242	1950	37.5	NORMAL	58	254	221	5,000	1650	34.5	NORMAL	49	225	195
2700	46	NORMAL	86	288	250	SL	2550	44	NORMAL	78	278	242	2250	40.5	NORMAL	65	260	226	1950	37.5	NORMAL	54	235	204	SL	1600	34.0	NORMAL	46	210	182

SPECIAL NOTES

(1) Make allowance for warm-up, take-off, and climb plus allowance for wind, reserve, and combat as required.
(2) High blower above heavy line.

EXAMPLE

At 11,000 lb gross weight with 280 gal of fuel (after deducting total allowances of 20 gal) to fly 1220 stat. air miles at 5000 ft altitude, maintain 1950 rpm and 37.5 in. manifold pressure with mixture set: NORMAL.

LEGEND

ALT - PRESSURE ALTITUDE
MP - MANIFOLD PRESSURE
GPH - US. GALLONS PER HOUR
TAS - TRUE AIRSPEED
KN - KNOTS
SL - SEA LEVEL
F.T. - FULL THROTTLE

REVISED 1-24-47
DATA AS OF 9-10-44
BASED ON: FLIGHT TEST

F-51D-1-93-6

FUEL GRADE: 100/130

FUEL DENSITY: 6.0 LB/GAL

Figure A-15. Flight Operation Instruction Chart—With External Load—12,200 to 10,300 Pounds

WADC Form 241K (11 Jun 51)	AIRCRAFT MODEL (S)	FLIGHT OPERATION INSTRUCTION CHART	EXTERNAL LOAD ITEMS
	F-51D ENGINE(S): (1) V-1650-7	STANDARD DAY CHART WEIGHT LIMITS: 10,300 TO 8100 POUNDS	TWO 1000 LB BOMBS, ONE 1000 LB BOMB PLUS CNE 110 GAL DROP TANK, TWO 110 GAL DROP TANKS, CR TEN 5 IN. ROCKETS

LIMITS	RPM	MP IN. HG	BLOWER POSITION	MIXTURE POSITION	TIME LIMIT	COOLANT TEMP	TOTAL GPH
WAR EMER	3000 3000	67 67	LOW HIGH	NORMAL NORMAL	5 MIN	135°C 135°C	210 210
MILITARY POWER	3000 3000	61 61	LOW HIGH	NORMAL NORMAL	15 MIN	135°C 135°C	180 180

INSTRUCTIONS FOR USING CHART: Select figure in FUEL column equal to or less than amount of fuel to be used for cruising(1). Move horizontally to right or left and select RANGE value equal to or greater than the statute or nautical air miles to be flown. Vertically below and opposite value nearest desired cruising altitude (ALT), read rpm, manifold pressure (MP), and MIXTURE setting required. Refer to corresponding column and altitude for new power settings when gross weight falls below limits of this chart.

NOTES: Column 1 is for emergency high-speed cruising only. Columns II, III, IV, and V give progressive increase in range at a sacrifice in speed. Air miles per gallon (MI/GAL) (no wind), gallons per hr (GPH), and true airspeed (TAS) are approximate values for reference. Range values are for an average airplane flying alone (no wind)(1).

COLUMN I		FUEL(1) US. GAL	COLUMN II		COLUMN III		COLUMN IV		FUEL(1) US. GAL	COLUMN V	
RANGE IN AIR MILES			RANGE IN AIR MILES		RANGE IN AIR MILES		RANGE IN AIR MILES			RANGE IN AIR MILES	
STATUTE	NAUTICAL		STATUTE	NAUTICAL	STATUTE	NAUTICAL	STATUTE	NAUTICAL		STATUTE	NAUTICAL
			SUBTRACT FUEL ALLOWANCES NOT AVAILABLE FOR CRUISING(1)								
1330	1155	400	1505	1310	1700	1480	1895	1650	400	1980	1720
1195	1040	360	1355	1180	1530	1330	1705	1480	360	1780	1545
1060	920	320	1205	1045	1360	1180	1515	1315	320	1585	1380
930	810	280	1050	910	1190	1035	1325	1150	280	1385	1205
795	690	240	900	780	1020	885	1135	985	240	1190	1035
665	580	200	750	650	850	740	945	820	200	990	860
530	460	160	600	520	680	590	755	655	160	790	685
400	350	120	450	390	510	445	565	490	120	595	515
265	230	80	300	260	340	295	380	330	80	395	345
135	115	40	150	130	170	150	190	165	40	200	175

MAXIMUM CONTINUOUS						PRESS. ALT FEET	(3.76 STAT. (3.27 NAUT.) MI/GAL)						(4.25 STAT. (3.69 NAUT.) MI/GAL)						(4.73 STAT. (4.11 NAUT.) MI/GAL)						PRESS. ALT FEET	MAXIMUM AIR RANGE					
RPM	MP IN.	MIX-TURE	TOTAL GPH	APPROX TAS MPH	APPROX TAS KN		RPM	MP IN.	MIX-TURE	TOTAL GPH	APPROX TAS MPH	APPROX TAS KN	RPM	MP IN.	MIX-TURE	TOTAL GPH	APPROX TAS MPH	APPROX TAS KN	RPM	MP IN.	MIX-TURE	TOTAL GPH	APPROX TAS MPH	APPROX TAS KN		RPM	MP IN.	MIX-TURE	TOTAL GPH	APPROX TAS MPH	APPROX TAS KN
						40,000																			40,000						
						35,000																			35,000						
SEE COLUMN II						30,000	2700	F.T.	NORMAL	96	383	333	2600	F.T.	NORMAL	87	368	320	2350	F.T.	NORMAL	69	330	287	30,000						
SEE COLUMN II						25,000	2700	46.0	NORMAL	98	372	323	2400	F.T.	NORMAL	82	349	303	2200	F.T.	NORMAL	67	318	276	25,000	2050	F.T.	NORMAL	58	290	252
SEE COLUMN II						20,000	2650	45.5	NORMAL	91	347	302	2300	40.5	NORMAL	77	325	282	2000	33.5	NORMAL	61	296	257	20,000	2100	F.T.	NORMAL	57	283	246
2700	46	NORMAL	103	351	305	15,000	2500	F.T.	NORMAL	89	335	291	2300	F.T.	NORMAL	73	311	270	2000	F.T.	NORMAL	59	279	242	15,000	1850	F.T.	NORMAL	53	261	227
2700	46	NORMAL	98	330	287	10,000	2400	42.0	NORMAL	81	310	269	2150	39.0	NORMAL	68	288	250	1750	35.5	NORMAL	54	259	234	10,000	1600	32.5	NORMAL	48	239	208
2700	46	NORMAL	91	308	268	5,000	2400	42.0	NORMAL	75	288	250	2100	39.0	NORMAL	65	267	232	1700	35.0	NORMAL	50	239	208	5,000	1600	30.5	NORMAL	44	215	187
2700	46	NORMAL	86	289	251	SL	2400	42.0	NORMAL	71	268	233	2050	38.5	NORMAL	57	246	214	1600	33.5	NORMAL	45	214	186	SL	SEE COLUMN IV					

SPECIAL NOTES

(1) Make allowance for warm-up, take-off, and climb plus allowance for wind, reserve, and combat as required.
(2) High blower above heavy line.

EXAMPLE

At 10,000 lb gross weight with 120 gal of fuel (after deducting total allowances of 20 gal) to fly 565 stat. air miles at 5000 ft altitude, maintain 1700 rpm and 35 in. manifold pressure with mixture set: NORMAL.

LEGEND

ALT - PRESSURE ALTITUDE
MP - MANIFOLD PRESSURE
GPH - US. GALLONS PER HOUR
TAS - TRUE AIRSPEED
KN - KNOTS
SL - SEA LEVEL
F.T. - FULL THROTTLE

REVISED 1-24-47
DATA AS OF: 9-10-44
BASED ON: FLIGHT TEST

F-51D-1-93-7

FUEL GRADE: 100/130

FUEL DENSITY: 6.0 LB/GAL

Figure A-16. Flight Operation Instruction Chart—With External Load—10,300 to 8100 Pounds

A

B

C

PAGE NUMBERS IN BOLDFACE DENOTE ILLUSTRATIONS

PAGE NUMBERS IN BOLDFACE DENOTE ILLUSTRATIONS

PAGE NUMBERS IN BOLDFACE DENOTE ILLUSTRATIONS

P

R

S

T

W

MODEL NA-73. The first Army Air Corps XP-51 (73-3101, AAF No. 41-038) was first painted in British camouflage and assigned number AG-348. Paint was removed and aircraft was delivered to Wright Field in August, 1941.

MODEL NA-73. The second of the Army Air Corps XP-51 airplanes. This is the 10th production airplane (c/n 73-3107, AAF No. 41-039). It was delivered to Wright Field, Dayton, Ohio, on December 16, 1941.

ATTENTION LUFTWAFFE! Keep away from this plane. Expect to see it on the farthest trip American bombers make. Expect to see it up high—40,000 feet—but don't expect to see it for long, because the Mustang travels at over 425 m. p. h. And you can expect to see more and more Mustangs, too. The men and women at North American are stepping up production every month. So when you see this high fighting, far flying Mustang, look out, Luftwaffe. Get out of there quick!

ATTENTION AMERICANS! BONDS bought these planes. WASTE FATS helped arm them. WASTE PAPER helped ship them. GASOLINE flies them. Will <u>YOU</u> help to deliver the next squadron?

North American P-51 Mustang Fighter

North American Aviation *Sets the Pace*

WE MAKE PLANES THAT MAKE HEADLINES . . . *the B-25 Mitchell bomber, AT-6 Texan combat trainer, P-51 Mustang fighter (A-36 fighter-bomber), and the B-24 Liberator bomber. North American Aviation, Inc. Member, Aircraft War Production Council, Inc.*

MODEL NA-91. There were 150 NA-91 airplanes built. Ninety-three were shipped to England as the Mustang IA. Shown above is P-51 (c/n 91-11982, AAF No. 41-37321). This was the second built for the Americans of this series and one of the 55 delivered. Mustang was "Apache" at this time.

MODEL P-51B. There were 1988 P-51B airplanes produced for the Army Air Corps as North American models NA-102 and NA-104. Through Lend-lease, 308 of these were re-assigned to the British as the Mustang III. Shown with American stars, British fin flash, camouflage and number (FX883) is one of these airplanes.

This Peace Talk Makes Sense

You don't have to translate the language of a Mustang's 50 calibre guns or the 75 millimeter cannon talk of a Mitchell bomber. The Japs already understand. So do the Nazis. So let's keep talking, You bet you can help! Think of it this way. The Bonds you buy help build these fast-talking planes. The waste fat you collect helps arm them. Waste paper helps ship them, and gasoline flies them. Wouldn't you like to say a few words of this kind of "peace talk," too?

THE MUSTANG IS A FIGHTER FROM THE GROUND UP! Speed: over 425 m.p.h. Ceiling: 40,000 feet Range: 2000 miles

North American Aviation *Sets the Pace*

PLANES THAT MAKE HEADLINES... *the B-25 Mitchell bomber, AT-6 Texan combat trainer, P-51 Mustang fighter (A-36 fighter-bomber), and the B-24 Liberator bomber. North American Aviation, Inc. Member, Aircraft War Production Council, Inc.*

MODEL NA-102. Prototype for the P-51D is this P-51B-1NA (c/n 102-24551, AAF No. 43-12102). This was the 10th P-51B airplane produced and was modified to the P-51D standards.

MODEL NA-104. Classic early morning side view shows P-51B-10NA (c/n 104-25144, AAF No. 43-7116) shortly after it was rolled off the production line.

SURPLUS MUSTANGS -

READ 'EM AND WEEP! The following Mustangs were sold in October, 1958 and their selling prices are shown in the boxes below.

The following description applies to aircraft Items 92 through, and including, 108. Aircraft are treated for storage, engines are pickled.

CODE 15a
AIRCRAFT, F-51D. The F-51D is a World War II, low wing, single tail, one place aircraft. The overall dimensions are: length is 32 ft 3-in., wing span is 37 ft, height is 12 ft 2-in., fuselage width is 2 ft 11-in., fuselage height is 6 ft 9¼-in. The F-51D is powered by one Rolls Royce design 12 cylinder, V type, 1400 horsepower rated engine, liquid cooled. Each engine is equipped with a Hamilton 4 blade hydromatic propeller. Aircraft is designed to use AN/ARC-3 radio receiver-transmitter or equal. The engine is manufactured by Packard Motor Company. The F-51D aircraft is manufactured by North American Aviation Company. Approximate net weight of aircraft is 7539 pounds. Tire size is 27-in. 10 Ply.

CONDITION: Used, minor repairs and usual work required to process aircraft for flight from storage.

ACQUISITION COST: $53,635.00.

92. F-51D Serial No. 44-11453 — 1 Each $________
Acceptance Date: 7/6/48
Total Airframe Hours: 1626:15
Engine, S/N V-333826
Total Hours: Unknown
Hours Since Overhaul: 144:55

93. F-51D Serial No. 45-11636N — 1 Each $________
Acceptance Date: 7/3/45
Total Airframe Hours: 1717:00
Engine, S/N V-335447
Total Hours: 784:10
Hours Since Overhaul: 74:40

94. F-51D Serial No. 44-72192N — 1 Each $________
Acceptance Date: 1/8/45
Total Airframe Hours: 1410:35
Engine, S/N V-331749
Total Hours: 702:20
Hours Since Overhaul: 140:20

95. F-51D Serial No. 44-73423N — 1 Each $________
Acceptance Date: 3/22/45
Total Airframe Hours: 1476:05
Engine, S/N V-323808
Total Hours: 343:40
Hours Since Overhaul: 59:50

96. F-51D Serial No. 44-75007N — 1 Each $________
Acceptance Date: 7/12/45
Total Airframe Hours: 1711:30
Engine, S/N V-320505
Total Hours: 411:35
Hours Since Overhaul: 9:10

TYPE AIRCRAFT	ITEM NO.	AMOUNT OF HIGH BID
F-51D	92	$ 1525.26
"	93	1499.99
"	94	1475.60
"	95	1525.01
"	96	1525.57
"	97	1450.57
"	98	1465.00
"	99	1400.00
"	100	1321.50
"	101	1449.89
"	102	1356.00
"	103	1451.58
"	104	1449.99
"	105	1551.01
"	106	1465.58
"	107	1551.02
"	108	1457.06

AVERAGE PRICES	
T-6G	$ 1,425.09
T-28A	2,917.98
C-45G	7,274.77
B-25	6,432.68
C-46A	41,003.42
B-26	3,447.33
F-51D	1,465.92

97. F-51D Serial No. 44-73825N — 1 Each $________
Acceptance Date: 4/10/45
Total Airframe Hours: 1321:25
Engine, S/N V-324467
Total Hours: 855:55
Hours Since Overhaul: 89:20

98. F-51D Serial No. 44-73343N — 1 Each $________
Acceptance Date: 3/14/45
Total Airframe Hours: 1405:10
Engine, S/N V-334202
Total Hours: 751:25
Hours Since Overhaul: 194:15

99. F-51D Serial No. 44-73210N — 1 Each $________
Acceptance Date: 3/6/45
Total Airframe Hours: 1357:25
Engine, S/N V-320017
Total Hours: 553:00
Hours Since Overhaul: 211:45

100. F-51D Serial No. 44-63889N — 1 Each $________
Acceptance Date: 8/12/44
Total Airframe Hours: 1477:44
Engine, S/N Unknown
Total Hours: Unknown
Hours Since Overhaul: Unknown

101. F-51D Serial No. 44-73240N — 1 Each $________
Acceptance Date: 3/7/45
Total Airframe Hours: 1359:05
Engine, S/N V-328031
Total Hours: 809:05
Hours Since Overhaul: 23:30

102. F-51D Serial No. 45-11540N — 1 Each $________
Acceptance Date: 7/8/45
Total Airframe Hours: 1383:35
Engine, S/N V-333818
Total Hours: 598:10
Hours Since Overhaul: 173:00

103. F-51D Serial No. 44-74589N — 1 Each $________
Acceptance Date: 6/12/45
Total Airframe Hours: 1910:20
Engine, S/N V-329510
Total Hours: 428:40
Hours Since Overhaul: 134:35

104. F-51D Serial No. 44-73543 — 1 Each $________
Acceptance Date: 3/23/45
Total Airframe Hours: 1628:45
Engine, S/N V-334393
Total Hours: 731:34
Hours Since Overhaul: 85:40

105. F-51D Serial No. 44-72936N — 1 Each $________
Acceptance Date: 2/20/45
Total Airframe Hours: 556:50
Engine, S/N V-329343
Total Hours: 1342:55
Hours Since Overhaul: 304:50

106. F-51D Serial No. 44-73320N — 1 Each $________
Acceptance Date: 3/10/45
Total Airframe Hours: 1182:40
Engine, S/N V-334995
Total Hours: 459:05
Hours Since Overhaul: 233:25

107. F-51D Serial No. 44-73518N — 1 Each $________
Acceptance Date: 3/23/45
Total Airframe Hours: 1194:00
Engine, S/N V-323244
Total Hours: 635:15
Hours Since Overhaul: 25:45

108. F-51D Serial No. 44-74960N — 1 Each $________
Acceptance Date: 7/14/45
Total Airframe Hours: 1203:35
Engine, S/N V-334634
Total Hours: 733:45
Hours Since Overhaul: 136:55

MODEL NA-103. There were a total of 1750 P-51C airplanes produced. Shown above is P-51C-10NT (c/n 103-26574, AAF No. 43-24943) one of 550 airplanes of this "dash" number. Note sliding cockpit window for cooling during engine run-up.

P-51D MUSTANG. Operational photograph taken in India in 1945 shows the familiar P-51D with dive-bomb markings on the wings. Later re-designated F-51D, this is the basic airplane described in the previous complete Pilot Handbook, the legendary "-1" of the military.

MODEL NA-109. The P-51D-5NA (c/n 109-27000, AAF No. 44-13366) was one of 800 "-5NA" produced in California. There were a total of 8502 P-51D airplanes built. California (6502), Texas (1600) and Australia (200). Including the P-51K (similar except for propeller) there were 10,002 airplanes of this basic version built - by far the most common.

MODEL NA-105A. Seven experimental light-weight Mustangs were built and tested by the AAF and the British. Shown here is the first of two XP-51G (c/n 105-25931, AAF No. 43-43335). Other light weight airplanes were the XP-51F (3 built) and the XP-51J (2 built).

MODEL NA-126. Photograph shows the 5th P-51H-1NA (c/n 126-37590, AAF No. 44-64164). Note that it has the short tail of the P-51D. The familiar tall tail of the P-51H was not added until the 33rd production airplane.

MUSTANG II. Cavalier Aircraft of Sarasota, Florida has been the most active in the conversion and modification of the postwar Mustangs. Shown here are two of the converted F-51D versions known as the Mustang II. Shown are USAF No. 67-14866 and 67-22580. Several of these were supplied to South Amrerican governments under the MAP as counter-insurgency fighters.

CIVILIAN MUSTANG. Completely rebuilt to "as new" standards by Trans Florida Aviation (now Cavalier Aircraft), the Mustang gained a new life as sport aircraft. Shown here as a two seater, with new instruments, avionics and wing tip tanks. Other firms also modified the Mustang for civilian use in the 1960's and 1970's.

TURBO MUSTANG III. Addition of a Rolls Royce Dart 510 engine of 1490 hp gives the long nose look to this Cavalier Mustang. This was built for the Department of Defense demand for an "AX" airplane. Installation was from a Viscount and included engine and cowling. A further version was built by Piper but not produced.